ALANA KAREN

THE ADVENTURES OF WOMEN IN TECH

HOW WE GOT HERE AND WHY WE STAY

Minneapolis

ISBN 13: 978-1-63489-381-7

Library of Congress Catalog Number: 2020917289

Printed in the United States of America

First Printing: 2021

25 24 23 22 21 5 4 3 2 1

Cover design by Luke Bird

Interior design by Patrick Maloney

Wise Ink Creative Publishing
807 Broadway St NE
Suite 46
Minneapolis, MN, 55413

To order, visit www.itascabooks.com or call 1-800-901-3480.

Reseller discounts available.

To my grandmothers, who loved to read and believed in lifelong learning, and my children, who are following in their footsteps

Contents

Introduction

I was born in 1977 when earlier feminism was very much in style. My mom did not like Barbies and forbade me to watch television shows with poorly written female characters. *Three's Company* sent her over the edge. The core storyline: Jack Tripper, played by the charming John Ritter, pretended to be gay so that he would be permitted to live with two women who wore short shorts. The landlord was really nosy. Comedy ensued. No, it hasn't aged well. Ritter's smile was killer, though—but I digress.

Many kids don't experience living away from their parents for the first time until the age of eighteen, when they head off to college and move into a dormitory. In my case, I was born and raised in one. My mom worked for Douglass College, the all-women campus of Rutgers University in New Jersey. She was an area coordinator, which was a weird hybrid of general manager, landlord, and dorm mom. My memories of her job run the gamut. She ran multiple dorms, which frequently meant dealing with dramatic fights, loud music, and burnt popcorn–induced fire alarms. She also organized a complicated manual process for the dorm room lottery each year, which involved a massive cork board and the use of many colors of paper. She was called in the middle of the night for attempted suicides and consulted with troubled students. She had to maintain files for each student and signed off on them every year in folders filled with dot matrix computer paper. She was working on that project the day I was born.

My mom's feminism was mighty and angry; it wasn't an option. According to the National Committee of Pay Equity, women earned 58.9 percent of what men did the year I was born. No wonder my mom kept working after her labor pains started. Salaries slowly rose through the 1980s, reaching 66 percent as compared to men by 1989. My mom was the primary income earner in my house throughout my childhood, and after twenty years of working at Rutgers, she made only $5,000 more than my starting salary of $35,000 at my first job at a start-up.

Growing up, I had the luxury of believing her feminism was funny. She often wore a bright red T-shirt she received at a conference that I called the "weeman, wyman, wooman" T-shirt. It listed all the various ways *women* had been spelled over the centuries. As a child, I thought this was ridiculous since most of the spellings were hilarious. Now I realize it was a tale of survival no matter what we were called.

By the time I was looking for a job at the age of twenty-two in 1999, I thought people had evolved beyond this inequality thing. After all, women were graduating college at a higher rate than men, so we were all caught up, right?[1] I was friends with men and women, both equally smart and funny, who went to the same school and were graduating with degrees. And now we were off to the workforce! If I had taken a moment, I would have noticed that more of the men were going into finance and computer science than the women.[2] I didn't ask whether more women were weeded out by certain courses than others. Or did men fare better in the on-campus job interviews for some companies than others? I didn't pay much attention at the time.

I naturally gravitated toward liberal arts classes and graduated with a history major. With my interests leaning away from the moneymakers of the era, I considered myself my own worst enemy. But one thing saved me: I was really into the internet. My father had introduced me to computers early in life, and I was comfortable around them and already a fast typer. I liked making websites, combining art and technology to make clear and beautiful content with the tap of my fingers. I taught myself HTML and by 1997 was convincing professors that I could turn in web projects instead of essays. I enjoyed internships where I could get my

---

1 Statista. 2020. "Number of bachelor's degrees earned in the United States from 1949/50 to 2028/29, by gender." Release date: March 2019. https://www.statista.com/statistics/185157/number-of-bachelor-degrees-by-gender-since-1950/.

2 Randal S. Olsen. 2014. "Percentage of Bachelor's degrees conferred to women, by major (1970-2012)." Posted: June 14, 2014. http://www.randalolson.com/2014/06/14/percentage-of-bachelors-degrees-conferred-to-women-by-major-1970-2012/.

hands dirty and fix their websites, and I talked myself into an internship at Darden Business School working on Adobe Flash designs even though I was technically underqualified. And that's where, in a dark room, with men typing away who knew more than me, like some secret club, I should have noticed that I was a woman.

But I didn't.

Being raised by equality-minded parents who were tough on me because I could always improve, I accepted criticism as part of life. I was also used to being underestimated. I stand proud at the cute size of five feet one inch and come across as nonthreatening. I was accustomed to having to prove myself and had already gotten used to surprising people with my wit and intelligence. Didn't everyone have to prove themselves?

But the men in that lab at Darden were closed off, always coding quietly in the darkened room. I had to pick up many skills on my own. In fact, the manager of the lab made it clear that no one was going to slow down for me; I had to earn my spot. As this was 1998, there weren't any accessible classes on Adobe Flash programming available. One dude semi-befriended me; I sat next to him and made myself hard to avoid. I looked over his shoulder and quietly kept learning.

The combined experience of internships and online projects garnered my first job as a webmaster at my alma mater, the University of Virginia. I again talked my way into the position, convincing my former professors that I could revamp and modernize the School of Arts & Sciences website. I sensed that the most important criteria to them, more critical than technical skill, was making a beautiful website in keeping with the values and aesthetics of Jefferson's university. I went all out showing them the possibilities of different hues of cream and solid content management, and I got the job.

When I started in the summer of 1999, the webmaster for the University of Virginia, also a woman, nearly lost her mind. She could not believe a history major with essentially zero experience had been hired. She wanted to be responsible for all the websites to make sure the approach was consistent across the university's portfolio, and now a twenty-two-year-old was thwarting her plan. Once again I had to prove myself, and I found that listening to people and relationship management were far more critical than my technical skills. When I completed the successful revamp of the site only months later, I'd managed to convince her and her colleagues that I was not a mistake and was, in fact, talented.

Did I ever once think, *If I were a man, I wouldn't have to convince them*? Nope. I had zero clue that was a thing. Doubting my abilities seemed logical. After all, I had followed the age-old philosophy of "Fake it until you make it," and I didn't look around and think anyone else had it easier than I did.

It's only been in recent years, as I reflected on my career and story, that I saw something painful. If I had been only slightly less confident, only slightly more hurt, I would have quit. What if I'd been more offended? If I'd thought that as a woman I couldn't or wouldn't be taken seriously or couldn't be myself? The successful and meaningful career I've had, in a *Sliding Doors*–type scenario, could have turned out very differently. If I had decided that the battle wasn't worth fighting or that the lack of welcome meant I didn't belong, I wouldn't be here with twenty years of experience in tech. It's the risk in the whole situation that I find painful. Women like me have persisted and gritted our way forward. But what if we hadn't? And what about the many who don't join tech? Or join but don't stay?

We are at a critical point in our society where the women of my generation can help the women of the next generation grow and blossom in tech. But we need to be here to help, and despite millions of women working in tech, hiring and retention numbers are abysmal. A common statistic quoted says 56 percent of technical women in their midlevel careers leave their organizations.[3] Tech companies routinely publish their diversity statistics with many working to reach even 25 percent of women in technical roles.[4] On top of that, minorities aren't choosing tech careers

3 National Center of Women in Technology. 2016. "Women in Tech: The Facts 2016 Update. Catherine Ashcraft, Brad McLain,and Elizabeth Eger. "https://www.ncwit.org/sites/default/files/resources/womenintech_facts_fullreport_05132016.pdf.

4 Harrison, Sara. 2019. "Five Years of Tech Diversity Reports—and Little Progress." *Wired*, October 1, 2019. https://www.wired.com/story/five-years-tech-diversity-reports-little-progress/.

even if qualified.[5] If I were starting my career today, would I choose tech based on those numbers? Would I choose tech after hearing the stories about bro culture? Would I choose tech if I thought work-life balance were a problem?

After eighteen years at Google, I've had an adventure or two. What began as a start-up gig in 2001 has turned into a meaningful, multidecade career where I've held four different jobs, encountered countless obstacles, and managed and mentored hundreds of talented team members while having a fulfilling personal life as a wife and mother. It's been challenging and rewarding, and even when I complain, I love it. Yet we don't hear these comforting stories of achievable success for women when we talk about tech, ones that didn't require moonshot product vision or CEO aspirations but did provide a path toward success and fulfillment. It hasn't been a perfect journey by any means, and yes, I've still been tempted to leave, but more often I've wanted to stay.

Stories matter; they are how we learn. While data can alert me to a problem, only speaking with people helps me find perspective. With this book, I bridge that gap with hundreds of stories from the successful women I've met in tech—who are they and what are their paths? What skills were most critical to their success? How many pursued leadership and why (or why not)? What were the obstacles faced and the often-unseen ways they navigated them? What support did they receive and from whom, and how would they advise others going forward? What were their hardest days and how did they surmount them? These questions aren't just for the workplace, but also for their personal lives.

It bothers me that not everyone thinks they belong in tech. It hurts that we aren't all having amazing career experiences in a world that believes in moonshots and possibility. How can I make a difference? I want to push on the boundaries we've created with the limited stories of who belongs in tech. By widening our view to the diversity already among us, to the women who are both totally normal but possess their own unique

---

5 Bui, Quoctrung, and Miller, Claire Cain. 2016. "Why Tech Degrees Are Not Putting More Blacks and Hispanics Into Tech Jobs." *The New York Times*, February 26, 2016. https://www.nytimes.com/2016/02/26/upshot/dont-blame-recruiting-pipeline-for-lack-of-diversity-in-tech.html.

superpowers, so we can all feel accepted. By talking about the various ways we navigate our careers, how we help others, the burdens we feel, and the opportunities we see. And, finally, by digging deeper into why women stay and why we leave, so we can find the best ways to support each other and the generations of talent to come.

THREE PARTS

When I imagined the readers of this book, initially it was women working at tech companies who are forging their path and interested in others' stories, like me. I wanted to support them in their journey and help them feel more confident in themselves as women in tech, especially if they question whether they belong in tech or are technical enough, whether they are really leaders, or the various other ways we can doubt ourselves. Then I thought of students deciding whether to embark on a career in tech after graduation, trying to understand if tech is a place they could belong. Finally, I pictured allies or those curious and wanting to read to see more diverse stories of women in tech.

With those audiences in mind, I've structured the book in three sections:

Part 1 digs into the diverse women in tech I have met and interviewed, all pursuing both technical and nontechnical roles in technology companies, or technical roles within other companies. I explore the many ways they came to tech and their different backgrounds. I assert how we have to be ourselves, and the unique identities and values the different women demonstrate. Lastly, I explore the different ambitions we have, far from the cookie-cutter definitions of success often pushed on us in society.

Part 2 then asks the seemingly logical question, "OK, what happens when we get to tech?" I poke at the pros and cons we find when we are representatives of a minority population. I pick apart the trap of needing to be liked and ruminate on why women might not help other women. I also explore what it means to have a career in tech while simultaneously building an outside life and family, a topic near and dear to my heart.

Part 3 gets more practical. I share tools women use to grow and further their careers. I discuss the importance of our personal champions, whether mentors, sponsors, community, or friends. I tease out the difference between surviving and thriving, and then I take that to its logical consequence: Why do we stay in tech, and why do we leave?

I end with a conclusion that ponders the next steps for us all to further women's careers in tech. Thank you for joining me on this journey of discovery and listening.

A NOTE ON METHODOLOGY

Biases in data are real, and I'll do my best to balance out some natural selection bias (most critically survivorship bias) by interviewing women of various technical and nontechnical backgrounds, levels of seniority, and status of employment. So while I started with women I knew or I'd worked with at one point in my career, I branched out from there via personal introductions and LinkedIn networking. When I reviewed the backgrounds of those represented in my interviews, interviewees had worked at more than sixty-five companies, start-ups, or institutions and represented thirty-five fields of jobs.

I intentionally sought to interview women with racial or other intersections, since I believe we should see tech match the bachelor's degree population,[6] and I want to portray their voices loudly and clearly. At times I also specifically sought out different perspectives, e.g. women who left tech, were at start-ups, or founded their own efforts. Where possible, I also augment our stories with data from research, books, and published articles. Note that for this book's scope, I primarily focused on US-based companies and employees.

Lastly, to provide consistency between interview subjects, I used sets of standard questions to structure conversations. The questions were based on career conversations I have with employees, inspired by the GROW model,[7] and begin with the simple question, "What did you study

---

6 National Center for Education Statistics. 2019. "Educational Attainment of Young Adults: Percentage of 25- to 29-year-olds with a bachelor's or higher degree, by race/ethnicity: 2000 and 2019" https://nces.ed.gov/programs/coe/indicator_caa.asp.

7 Performance Consultants (International) Ltd. "The Grow Model." 2019. performanceconsultants.com/grow-model.

in school?"[8] In gathering quotes for the book, I was careful to only clean up language to aid understanding (e.g. cutting down repetition, removing crutch words such as *like, um,* and *you know*).

I will acknowledge, however, that Google is heavily represented in the background of my interviewees, as either a current or ex-employer. Over 60 percent worked there at some point in their careers, including internships. This is partially because I work there, and I have access to interesting women and they to me, but also due to Google's age. Even when I intentionally sought out others, I would often find they'd worked there at some point in their careers. Since Google regularly earns spots on lists for best places to work, that could have impacted the experience of these women. That acknowledged, I did hear differing experiences regardless of where women worked because their roles, managers, teams, and personal experiences varied. The majority of women also had careers spanning multiple companies, meaning their stories often included more than one place.

Finally, I am presenting this information as stories, as if I'd opened up my living room or office, and you got to hear what I hear daily. As a result, it's more anecdotal; by its nature, this work is not an analytical study. What I've captured here is a collection of what we sense versus hard data. Sense is how we feel, what we observe, how we perceive our experiences, and what we take away. That's what I heard when I was interviewing for this book and what I want to share. Because, whether we like it or not, data—which I'll quote at times—is only part of how we make decisions. More often than not, we are propelled by sense.

You also won't find me challenging or debunking these women during their interviews. I did the opposite: I was grateful for their stories and accepted them as their truth. And for the most part, I heard how each woman felt, what she observed, and what she remembered on a particular day, which also influenced what I learned. I took my role in listening to these women, and their trust in me to share their stories, very seriously.

---

8 Schneider, Michael. 2018. "Google Managers Use This Simple Framework to Coach Employees." *Inc.*, July 30, 2018. https://www.inc.com/michael-schneider/google-discovered-top-trait-of-its-most-effective-managers-you-can-develop-it-too.html.

Where requested or appropriate, I anonymized their identities with pseudonyms or aggregated data for additional comfort.

You may read this book and say to yourself, "That's not true!" based on your own experience. You may be angry or disagree with the fact that I've included a particular example or quotation. One lesson I've learned in my career is to listen to learn and not to fix. While this is semifunnily captured in jokes about how women just want others to listen and not to fix them, it's a true way of listening for all of us. Can we take these stories at face value and seek to understand someone else's experience? If this is where they are, how do we positively influence their careers?

My ask? Read these stories with a curious mind and think about possibilities, for ourselves and others.

PART ONE

YOU BELONG IN TECH

"I think it's great that you're doing this. What I always think of at this stage in my career, particularly as a breadwinning woman, is, twelve years ago when I was in this position, if I'd had a woman twelve years ahead of me that could be like, 'You're going to be fine,' that would've been amazing. And that's what this project is. This is what it looks like, and this is how it goes for the vast majority of us. And I think it's really important to showcase that and to give it a voice in a community. Because I want to be able to look at a woman who's in her midfifties and . . . she's reinventing herself, or she's still crushing it and enjoying still being in this crazy rat race, right?"

—Bethanie Baynes

1 | We Come from Everywhere

MY BARELY THERE THESIS

I knew early on I was going to title this chapter "We Come from Everywhere." With news headlines hitting every day about tech's diversity challenges,[9] I might appear to be taking a controversial stand. It doesn't feel that way to me. After twenty years in tech roles, I can confidently say I've lived among diversity. I have met people from vastly different backgrounds than mine, and I have seen us support each other despite our differences. Should there be *more* diversity? Of course. An emphatic *yes*. With that, we also need to recognize and accept the different perspectives and backgrounds in our community, or we're overlooking a key tenet of inclusion: *making each of us feel like we belong*. And if we fail there, we risk retaining the diversity we do have. And so I've focused this chapter on how we are unique and need those differences.

Why is this important? Diversity of ideas, experiences, and points of view is critical. More and more research backs up the importance of this, not only for how we feel but also for bottom-line performance and economic results. A National Center for Women in Technology infographic boils it down nicely.[10] In a study of 500 US-based companies, higher levels of racial and gender diversity were linked with increased sales revenues, market share, and relative profits as well as more customers. In another study, teams with equal numbers of men and women, as compared

9 Bogost, Ian. 2019. "The Problem with Diversity in Computing." *The Atlantic*, June 25, 2019. https://www.theatlantic.com/technology/archive/2019/06/tech-computers-are-bigger-problem-diversity/592456/.

10 National Center for Women & Information Technology. 2016. "Women in IT: The Facts Infographic [2016 Update]". Accessed May 27, 2020. https://www.ncwit.org/resources/women-it-facts-infographic-2016-update.

to teams of other compositions, were more likely to be creative, experiment, share knowledge, and complete tasks. Interestingly and somewhat counterintuitively, a series of studies using mathematical and computer modeling found that diverse teams consistently outperform even teams made up of the highest-ability performers. The strongest teams have differences in opinion. Without that, we risk a lack of critical thought that leads to failed ideas and products.

As we've begun to expose diversity issues in tech, we've tended to focus on specific problem areas. This is influenced by what tech companies share in their diversity reports, which focus on race and gender. This is critical, of course, and I am thrilled we are shedding light on important issues and working hard to resolve them. You'll see me highlight those issues through the book. That said, I will also share stories of women with other differences so we can see many kinds of careers, ambitions, and backgrounds represented, like geographical locations, socioeconomic statuses, and life experiences. Why? If we only tell the stories of geeky engineers, tech bros, and suited MBAs, we ignore and silence the richness of the community. We're yet again failing at inclusion where it matters most: recognizing and acknowledging each other as a part of the whole. So let's change that now.

In my interviews I was struck by the huge variety of backgrounds that led women into their tech careers. One's love of numbers led to accounting, then advertising and digital media attribution. Another woman's customer service gig led to communications, writing, and management opportunities. The stories go on. This chapter will give you a flavor of what inspired me to write this book.

To give you a sense of what I learned, I've summarized profiles of different women, their current roles, and how they found their way to tech. I don't cover every woman I talked to in this chapter, and beyond that there are so many women I didn't speak with who would encompass even more experiences, paths, and backgrounds. You'll hear more from the women introduced throughout the book, and I'll discuss other women as well. There are so many that I don't expect you to remember and keep track of each woman. Instead I suggest you see what stories surprise, resonate with, or stick with you.

ORIGINAL DREAMS

It's worth noting that, first and foremost, we got lucky. Most women did not grow up thinking they wanted to work in technology companies. There are many reasons for this, some obvious—like how the internet triggered new innovation that didn't exist when we were younger—and some less so. I have a BA in history, and I also studied French and creative writing. Does that sound like the background of someone who's worked at Google for eighteen years? It should, as it turns out. During my interviews, I saw a pattern of women ending up somewhere very different from where they began.

Annie Lange, Technical Program Management

Annie "imagined being in a cabin in Maine writing books, which is the opposite of what I'm doing now, which is cross functionally working with everybody and organizing their lives or what they're doing." After graduate school, she started working as a digital librarian and considered pursuing a PhD in poetry. On a trip out to San Francisco, she visited a friend working for a thirty-person start-up. She was immediately struck by the "intelligent, motivated people, and they're moving fast, doing things and iterating, and you don't have to wait months to get reviewed by a committee to make one change."

She did not take her flight home, instead quitting her job remotely and joining an early-stage start-up. She was thrown into working with engineers and figuring out how to navigate on her own. When the start-up was acquired by Salesforce, she became an engineering manager based on her work at the time.

Camie Hackson, Software Engineering

Camie always thought she was going to be a doctor. She took pre-med classes at Berkeley, but the university didn't have a pre-med major, so she chose computer science (CS), which intrigued her. She took a gap year after graduating to work while she applied to medical school. "It was really easy to get a job in Silicon Valley with my CS degree. And then six weeks into working in Silicon Valley, I realized there was no way I was going to med school; I was having too much fun."

Ashley Sun, Software Engineering

As Ashley was taking AP bio in her junior year of high school on her then-presumed path to medical school, she realized something. She "absolutely hated it." She applied to various colleges for various majors, and she "happened to apply to UC Berkeley as they have an undeclared engineering major." That path ultimately led her to computer science courses. One of her first was a Matlab course she really liked. "I thought the concepts and the way of thinking [were] really cool and different. I did really terribly in the course. I still really liked it." She decided to pursue an electrical engineering and computer science degree (EECS), "which sounds way fancier than it is, but you really only have to take two or three electrical engineering courses and all the rest are CS."

While it was a tough road to start taking CS at this point in college when other students had been exposed to coding earlier, she endured. She interned at LendingClub one summer, and they asked her to come back full-time after graduation. "I feel like it was good luck." In 2019 she took a break to travel the world before joining a new company.

Kathleen Fletes, Human Resources

Kathleen was inspired by art and performance, and she wanted to be a Broadway actress, movie producer, or movie director. "I think that was a bit short lived. It was just 'because I liked cinema and I liked the arts.'" In late high school, she was really good at math, and people told her she should go into engineering. "I actually started my major in civil engineering and then got my first C in math, and it was devastating. And I was like, *I can't do this!* I ended up switching my major." Since graduating college, she's always worked in human resources (HR), first as a temporary worker and then full-time.

Sara Phillips, Learning and Development

"I didn't actually go to school. After high school I had moved to Los Angeles to act and had a great year there. I didn't do very much acting but went to some amazing parties in Malibu." Sara returned home as tech was taking off. "I think that my entry into tech was serendipitous. I was starting my career at the time of the next wave of the tech boom after dot-com [bust]." She started off doing phone support at the customer support center for Boost Mobile, and she worked her way up. "I wound up managing a team of supervisors and a quality program. It just so happened to coincide with the launch of the first iPhone; we ran a successful pilot for Apple, and we got to provide support for the very first iPhone, and it was away from there."

WHEN WE KNEW

As educational institutions evolve and integrate technology into their offerings, prospective employees can learn relevant skills before joining companies. This means we can now target tech as our ambition earlier in our careers and collect applicable knowledge to get us hired, or also further ourselves through additional tech education later. We are far from the days when I had to teach myself HTML! Tech has also grown into a significant industry with career offerings for people from various backgrounds, not only technical ones. Taking these trends together, people may now decide tech is the right place for them at various times throughout their career.

Reese Pecot, Trust and Safety

Reese originally thought she'd be an animal rights attorney, and then she became more interested in the political environment as she grew older. "It was in law school that I realized that a lot of what was happening in our world was going to be impacted by technology." That realization spurred her journey into tech, first as an intellectual property lawyer and then as a director in trust and safety. "It was very much in order to figure out how to use tech to make the world a better place."

Paola Nash, Program Management

Born in San Francisco, Paola describes her career in tech starting "by osmosis." Amid all the tech companies popping up, she remembers "one of my first interviews as a senior in college was for Facebook in a user operations specialist role." Being the first generation in her family to graduate from college, she felt pressure to find a job graduating amid the 2008 financial crisis.

"I didn't have a financial safety net to take my time finding a role in line with my interests, passions, etc.; all those uplifting keywords used in commencement speeches. I was in survivor mode along with other folks at that time. As a first-generation college graduate with a newly minted UC Berkeley degree, the pressure was on me to succeed and not disappoint. Whatever companies were hiring, I was going to try to get my foot in the door." Garnering her first role in tech through a contract customer service gig, Paola has sought ways to grow and expand her "roots in the ground" into her current career in program management.

Kris Politopoulos, Technical and Engineering Operations

Kris had a technical knack from an early age, when she would program in Basic on her home computer. She loved figuring out how things worked, and she ended up being technical support for many of her friends in her personal life. Never professionally trained, while doing an accounting job she started to call in favors from friends who were in technical roles. "You got to get me a foot in the door, and you know I can do it. I'm smart enough to do whatever." Her future husband was the person who gave her a chance, and she started doing desktop support at a clothing manufacturer. She converted every job she had into a more technical role, performing IT support and intranet management. By 2002 she was able to move into a fully technical job.

Marily Nika, Product Manager

Marily didn't know what she wanted to be when she grew up, but she knew it would involve computer science. "All I knew was that I wanted to pursue a career that involved working with computers and getting my hands dirty by spending hours solving puzzles and problems on them. I would enjoy practical work way more than reading a book on history or literature. I was really creative at plugging computer science into any homework I was assigned, whether this was economics (writing scripts for the equations), literature (using my computer for research/essays instead of heading to the library) or even art (learning photo editing software from a young age). I grew up being surrounded by tech, and I loved it with everything I had—I knew that I wanted to pursue a career in tech ever since I was ten."

Natalia Lizon, Go-to-Market Team, Strategy and Operations

Natalia studied industrial management engineering and then got her MBA. "My engineering is less technical than traditional engineering and is considered to be more of a blend of art and science, including subjects like behavioral studies. It teaches a robust toolkit in terms of problem-solving, analytical thinking, modeling, and supply chain management. But it's a lot more oriented towards management consulting, for example, or roles that benefit from generic problem-solving, versus roles in coding or software development." During her undergraduate studies she identified part-time consulting engagements in "health care and nuclear energy around automation and digitization." While doing her MBA abroad, she applied for an internship at a tech company. "And then I guess the rest is history."

SEEING THE FUTURE

With the world rapidly changing, many women saw change coming to their industries. They were ahead of the game and wanted to dive in headfirst. Tech afforded them the opportunity to move fast and learn quickly in a new space. As you'll see, there was no formula to this—one woman was working on a farm, another in photography, another in childcare. What's similar here is the foresight of how tech would change the world.

Bethanie Baynes, Business Development

"So the truth is I read an article that Google gave away free ice cream on Fridays, and I was like, *This company sounds awesome.* I sent in my resume, and they called me back."

It was 2003, and Bethanie was working in the photo industry. The CEO for the company she worked for thought the internet was a fad. She moved to a smaller, artsy photo shop that had a website that was advanced for the time, with digitized files that were listed on the site. However, they weren't selling through the website yet. "And I was like, 'You're missing it. You're almost there, but you're still missing this link of where the world is headed.' So when I saw a company like Google—and to be honest, I really didn't know what they did—it piqued my interest that this is a company that's on the forefront of technology to the point where I can't really understand what they do. And that was really intriguing to me, and that was 2004, and then I flew out for my interviews and joined the [temporary workers] team and started approving ads all day, every day."

Camille Gennaio, Real Estate and Workplace Services

Camille was originally focused on actuarial science in school. "I was really good at math, and so my actual first major was actuarial science. I was given a proof at Howard University of why zero equals zero. And I just looked at the professor. I was like, 'No, I can't do this anymore.'"

Camille decided to change her major to education. When she was younger, if she was sick and unable to go to school, her working mother was forced to bring Camille to the office with her, as she had no alternate options. Camille had an idea that companies should provide childcare, and she used to question her mom's boss—"Why don't you have a better way for [my mom] to take care of me if you want her to work?" Then she would make him entertain her. She received a master's in public education, and the first company she worked for was in backup childcare. Coming "full circle," as she puts it.

"I saw Larry [Page] and Sergey [Brin] on TV, one of their early interviews. They were bouncing around on bouncy balls (disclosure: we're all the same age). And I laughed at the TV, going, 'When they're old enough to have kids, that company will have childcare.' And I wrote it down in a journal." The experience of seeing the Google founders stuck with her. As her visa from a job in Japan to open childcare centers for an international company was expiring, she looked online and discovered Google was hiring a hundred teachers. "They had decided to open these childcare centers. So I had told the future, and I applied and started in tech in that way."

Laura Kendall, Marketing

Laura studied at the University of Wisconsin–Madison. She attended the School of Business with a focus in marketing, which is also her current field, but also got a certificate in Spanish studies and studied abroad in Sevilla, Spain, her junior year.

"Early on I wanted to be a nutritionist or a large animal vet . . . but I also wanted to do something with computers. I grew up on a dairy farm in Wisconsin and got a *very* early start in databases. I would help my farmer father enter data into an on-premise herd management system . . . this later grew into a love for data and analytics and how they bring actionable insights into our everyday lives. But I also loved being around the cattle, the abundance of life, and growth around the farm, especially when the vet came and I could ask him every question under the sun (probably annoying him in the process). As I got older and learned more about the world through my college classes, I realized I wanted to do something with business, as I felt that was a way to have a greater impact. Marketing seemed like the best fit as a way to be persuasive and have impact."

Jill Szuchmacher, VP Operations

Jill was working in media at MTV to pay the bills after college, and she had a theater company that "did experimental multimedia, large-scale work where you had a controlling video and lights and sound." Each required its own discrete system, e.g., playing videos or DVDs. "So during a show that I was a stage manager for, as I was calling all those various cues, we had rigged up a machine that managed all the cues of this small, very complicated show. And it actually caught on fire in the middle of the show." She called the rest of the cues manually and then collapsed when the show ended. "I said, 'There's got to be a better way to do this with computers.' That was literally what I said. It was in 1998, I think."

Jill began her own software company to address this need, and she took it through the start-up stage before joining a tech company in business development.

BOLD MOVES

I was also impressed by some of the big leaps that led women to tech. Whether by forwarding their resume to broad groups of people, boldly selling their skills, or jumping into the unknown, many women ended up in tech roles by virtue of their grit, risk-taking, and good old-fashioned spunk. I'd like to think I rose to the same level; I did talk myself into my first job in tech, after all. Then I moved across the country to join a startup! Why is this important? We'll talk more about that in chapter 8, "Our Magic Toolbox."

Caragh Lavoie, Staffing Manager

"I had moved out to San Diego, California. It was going to be a year off with my friends from high school. And I had sent my resume to all the Villanova alum in the San Diego area. One of them was a president who owned a small executive search firm. He hired me, and then I started in recruiting. The majority of my early career was recruiting and [working as] a staffing manager but within the health care industry. And then I started with Schering Plough, which is a big pharmaceutical company." When Merck acquired Schering Plough, she was laid off due to duplicative teams. That's when she learned her current company was hiring for recruiters. She applied and started as a temp, which was standard for recruiting at the time.

Kristen Morrisey Thiede, chief people and business development officer at a startup

Kristen went to university at a small school in Tennessee and initially wanted to be a lawyer. During college, she had an internship at a television station in the advertising department. "So when I got out of college, I got a job at an advertising agency buying media—TV, radio, and packages like at the baseball stadiums." A vendor introduced her to an online marketing firm, which led to her next role buying online media, including from Google. Every time she called them to buy ads and "give them more money," it struck her that they had a growing ad sales team. "And so finally, the third time I talked with [their team], I was like, 'If you're hiring all these people, you should hire me.'"

She traveled from Atlanta to interview. "They flew me to New York, I had ten interviews, then I went home. I didn't get the job. And then two or three months later they called and said, 'Hey, are you still interested in the job?' And I said yes. And so they flew me back to New York and I did another ten interviews, and I flew back home and I called and said, 'Did I get the job?' And they said, 'No, you didn't get the job.' And so two or three months later I called them and I said, 'Hey, did you ever fill that job?' And they said, 'No, are you still interested?' I said, 'Yeah, I'm really interested.' So then I showed up with a PowerPoint presentation on how I would do the job and how Google couldn't survive without me, so then they hired me."

Hillary Frank, head of marketing and community at a venture capital and private equity company

Hillary studied retailing and consumer studies, a major that was officially in the School of Agriculture at the University of Arizona. "I liked the idea of being in a business that was about . . . consumerism. What does it mean when you walk into an experience in a store online? It was just coming up in school; I graduated in '98, so e-commerce was on the cusp of happening."

She decided it was time to take the leap and leave Arizona, where she'd been born and raised. At the time she didn't seriously consider a tech company. "I am not technical in any way; I didn't actually translate that I could work in that area because I'm not a technical person. And I thought you had to be a coder." She took a job with a company at their corporate headquarters in the Bay Area, hoping it would lead to something bigger over time. Afterwards she moved into a recruiting role supporting technical companies and later joined a start-up as employee number ten, focused on customer service. That role in turn led to a Google role, which was a hard decision given how well she was doing at the start-up. "But

I knew that Google would completely change the course of my career, which it did. And I am forever grateful for that."

Tieisha Smith, VP, Regional Technology Manager

Tieisha has been in the technology industry for over twenty years, but she "fumbled around to find my niche." She initially went to college for computer science, but she had "no passion for it." She pursued a major in marketing and then a graduate school MBA in high technology management. The ultimate learner, she started her career at the help desk, then moved to testing support applications, before trying out a business analyst role. When her role was outsourced, she joined Marsh & McLennan initially in a support role which turned into a people management gig. Now there for sixteen years, she's relished the opportunity to try different areas, earn new certifications, work with the different subsidiaries, and become an expert in change management. Her exploration has led her to different opportunities like her current regional technology manager role but also different states like Arizona and California.

THE INSPIRATION

Throughout this book, I've been struck by how meeting the right person at the right time can make a difference in our careers. Whether it's through coaching, mentoring, or just being on the stage as an example, our careers can be shaped by who we listen to and who we meet. In the "Our Champions" chapter, I'll talk about this more. Here's a quick flavor.

Jessica Taylor, Marketing and Start-Up Founder

Jessica was an English and French major at the University of Virginia. After working for seven years "for some start-ups that were poorly run by previous English majors that knew nothing about business," she decided to go to business school. She chose to attend the College of William & Mary strategically, as she had a two-year-old at the time and didn't want to take on debt. "I figured I'd get an MBA at a regional school and then probably end up working in Richmond and living there close to my family in Charlottesville."

Her fate shifted after attending a conference as the president of the National Association of Women MBAs. "Kim Scott was the keynote speaker in a room . . . it was actually a church, and there were a thousand women there or something. I think her parents were there; it was a big deal. This was in 2006. And she had a pretty feminist message, and I really liked it and I really liked her style." At the time, Kim Scott worked at Google on the AdSense publishing team. And so Jessica sent an email to them. "It was ridiculous. It was like a paragraph and it was like, 'If you're looking for a spunky woman MBA, I'm your gal.' And I attached my William & Mary resume." A few weeks later, she received a call from their recruiting department.

Belvia Sharp, Executive Business Partner

Belvia always tells people, "I am not your typical tech person. I grew up in middle-class America, but I didn't go to the fancy schools, then have aspirations to go to some Ivy League college and then set off to do XYZ." Her first love was fashion design, which her high school home economics teacher helped encourage. "If I didn't have that for school, I probably would've been a high school dropout because I hated school." She'd learned how to use sewing patterns from her grandmother, and her teacher said she had a knack and should go to school for fashion. But her mother felt the opposite. "With my mother growing up in the Jim Crow days, everything was black or white in the house. So it was either black folks didn't get to do that or, no, you need to go work for the state where you have that steady kind of job."

Belvia still decided to go to fashion design school and also started working at Britex Fabrics, where she took charge of the bridal department. But she hated it. "I actually loved sewing for people. I just didn't want to be in retail anymore. I was there because I was going to school. But then I had my son." She was married and had another mouth to feed. In the summer of 1992, she realized she didn't want to keep the Britex job, but her husband said they needed the income. After a month of getting in the car and saying she was going to quit, "I got in the car one day and I said, 'Yeah, I've got two weeks left.'"

Her husband was learning DOS (command line prompts for IBM

personal computers), but Belvia had no enthusiasm for computers, which looked "foreign" to her. But her husband kept harping at her, and she signed up at a temp agency in San Mateo, California, saying she was willing to do anything, to work hard, and to learn. "This was Wednesday. Thursday [my contact at the agency] called me and she's like, 'I found something for you; someone's going on vacation, but they said they'll train you and it's only two weeks.' And that was my entry into Electronic Arts."

Sarah Milligan, Communications Manager

Sarah received her BA in journalism with a minor in criminal justice from the University of Nevada. She says her career in tech started by accident. "I was working as a communications specialist at a small consulting firm, and my manager pulled me aside and said something along the lines of, 'You have a lot of potential, and this is a small company where you don't have room to grow.' His wife, a chief of staff at Microsoft at the time, was looking for a contractor to manage executive communications for the Americas operation. I was twenty-three and had a new baby." The role paid almost twice what she was making at the consulting firm but felt risky because it was only a contract position. But despite feeling "severely underqualified," she thrived.

Wendy Zenone, Security Engineering

Wendy grew up with a stay-at-home mom and thought she'd take that path too, realizing later it wasn't the life she was looking for. Having not attended college, "I went back to school and got my esthetician license because I had a son at this point. I was twenty-five." Initially thinking she'd be set with that career, she realized it still didn't cover the bills. She ran the numbers and applied to college, securing grants and scholarships to study communications. An internship under a strong mentor led to her first full-time job in PR. Transitioning to tech with an entry-level ads job at Facebook, she became interested in information security and reached out to a female colleague in the field. Inspired by what she saw, she pursued social media and communications jobs in that space. With the support of her then-boss, she applied to Hackbright Academy, a selective all-women's ten-week software engineering program covering the technical skills needed to become a full-stack software engineer. This ultimately led to her current career in security engineering at Netflix.

WHY TECH?

Despite our differences, there is one area where I found that women in tech are extremely alike. During interviews, I asked the women what they liked about working in tech. As if they had rehearsed their responses, nearly all of them shared similar thoughts: we love the speed of innovation, the ability and scope to change the world, the career opportunities, and the people. Is the story so identical because we've been brainwashed? Possibly, but I think it's more likely self-selection. Those of us who were drawn to tech, and who have stayed, chose to come for these reasons and have chosen to stay for those same reasons, even when struggles arise.

The Love of Speed

Don't get me wrong, I love stability. I essentially stayed in my first job at Google for ten years while others were typically moving on every two to three years. However, I stayed so long because new and interesting problems to solve were always coming up. Many of the women I spoke with feel similarly, as indicated by these example quotations from our conversations:

"I love that we move so quickly. Every day feels like something new, a new problem to solve. I also feel like we're solving the sort of big challenges that haven't been tackled before."

"There's always going to be something to do, and there's always going to be something interesting to do. It's not going to be 'turn the old wheel' of whatever. Actually, in six months, just do something completely different if you choose, or be exposed to something completely different . . . which is fascinating, exciting, [and] a little bit terrifying."

"I like building, I like solving problems. I think that's where I get my energy. . . . I've been able to reinvent my career a number of different times based on the fact that we're growing and changing. And there are new problems to solve."

Times Are Changing

Paired with speed, tech also has an enduring openness to change. This likely stems from its original research origins, which bred an environment open to arguing for the best ideas and striving for continuous innovation. This applies not only to technology but also to the business processes and policies:

"I don't know if I would do well in some place where you just came in every day and had to lead a staffing team . . . [or] you just had to hire folks and you weren't trying to think through how we have a more representative [workforce], or how we are thinking about internal mobility, or how we are thinking about knowing our internal talent. So I like the fact that it's a place where we can still ask questions and want to do better."

"I've been here long enough that I'm like, it's always changing and it's fascinating, the amount of change and the quickness. As a person who does not like to get bored, it's the perfect place for me to be. . . . When I took the job, I was like, this is a company, but it's also a little mini universe of opportunity because, assuming success, it's going to grow and it's even more than I could have assumed or imagined."

The Chance to Change the World

Shaping the world has provided a significant motivator for bringing different types of people to tech. A number of women spoke about the power to help the world that they've felt and been able to harness regardless of their role.

Sherry Lin studied economics, what she calls the "major for lost children," at Georgetown. All she knew was that she wanted to see the world. Having graduated into the recession around 2010 with limited job opportunities, she felt free to do something "completely off the normal path, which is how I ended up moving to Hong Kong and then Burma."

She lived in Burma for two years, and she saw the country change quickly during that time. "So SIM cards used to be $800 apiece. And when I moved there, my office had to front me the cash for it, because I didn't have $800. And when I got the SIM card, I had no one to call because none of my friends had $800 to pay for the SIM card. While I was there, the government approved two multinational and telecom companies, and the price dropped overnight. It was $800 and then it was $5, and all of a sudden everyone in the country would find SIM cards for the first time." That quickly brought change to her work environment. She also saw her neighbor watch her first YouTube video, which changed Sherry's aspirations. "I was working with farmers at the time in a social enterprise, and it was to help the country. I was like, I've been working very hard trying to have a certain amount of impact, but the impact of this piece of technology dwarfed that." Having had that epiphany, Sherry moved to the Bay Area without a job, getting her first job in tech in a logistics role.

Sara W. thinks about how we can help as the world rapidly changes. "I think it's really interesting to think about both the good and the bad of what technology has done over the past twenty or thirty years in terms of how it's spread information and how that has led to a cultural transformation." How can we influence the cultural transformation and get more good stuff and less bad? "The flip side is the reality of the cultural transformation you want, when things are moving in a much faster way and it's really hard for people to keep track of." As both a writer and a tech employee, she's interested from a job perspective and storytelling one. "The stories we tell about who we are and how we think, like our vocabulary, has shifted a lot faster. Or like the things that the public talks about in public discourse, [which] seem to shift overnight. If you're not on Twitter the day that the blue and black dress thing happens, you miss it and you will never see it again." The speed and intensity of this change to how people consume information intrigues her. "Those problems are really interesting to me, and how you help humans that are not built for this speed of change navigate this world is really fascinating."

The Culture and People

I've realized over the years that the people you work alongside make more of a difference to your daily experience than the exact tasks you're completing. I often say culture is what we do every day. In other words, it's how we treat each other, what we say, and what we do relentlessly each day that make up what we perceive as culture. Many women saw tech's culture as open, honest, and intellectual, and that was a primary draw for them.

Camie was attracted to the meritocracy. Having volunteered at hospitals and doctors' offices and seen more traditional hierarchical environments, she was inspired by the culture of tech. "I felt like I had a voice in designing the products, and that it didn't seem to have the icky hierarchy that I felt that I didn't like in hospitals."

Tequila loves that "everyone shows up, but there are people that just show up whether you got on flip-flops and some shorts or you want to wear jeans and a cute blouse, but you show up and you're very honest about who you are." She loves that people can be their authentic selves. "I love being in a place where people can really show their true colors, and there's not a lot of people judging you."

From another longtime tech employee: "To this day, I do like that there are the brightest people that I probably will ever work with. I realized if I went to any other company or even different industries, you're not going to get the pace and the intellectual challenge and also the ability to collaborate with folks that you know are brilliant."

Practicality Rules

It's worth noting that there is a bottom line, especially depending on a person's finances, goals, and situation. Since tech thrived over the last few decades, women can be drawn in for practical reasons like job opportunity, flexibility, high salaries compared to other industries, and benefits. We shouldn't shy away from this or consider this shallow. The very diversity we search for comes across backgrounds and will make this a real factor in our hiring and retention of talent:

"Very randomly I was at a nonprofit, and then [my current company] came and they had a picnic at this nonprofit. They spent either $15,000 or $20,000 for renting the site for just that day. And then on that day I applied for every opening [they] had available from VP to manager."

"Yeah, I love it. I love the flexibility. When I think about my future and why I would be hesitant to get out of tech, it's because it is such a flexible culture. At least in my experience and most of my peers' experiences—

working remotely, running out of the office when you need to. It's not this rigid nine-to-five, coat-and-tie environment."

"I think there's a lot of opportunity here. There are a lot of jobs. You make a good living because a lot of these companies are competing for talent. There are rarely times where I have worried about what my family will eat. . . . That's a real luxury, I think. And if you find yourself at the right place, I think there's an element of luck that can really make your financial future."

Being Part of History

Related to the themes of impact and changing the world, there's also a sense among women of the role they are playing in history—the history of women, tech, and the world.

Reese speaks of this poignantly. "I feel [like] part of a historical moment. I say this even to my daughters—think about [the fact that] the first internet domain was registered in 1985. So we're in this period of rapid technological change, and to get to be part of that I think is tremendous. And so it does feel like living history, and it does feel like our daughters will look and say, 'Wow, my mom was part of the tech revolution.' And so, to me that's amazing, and it's also amazing that mothers are getting to say that and be part of that, and to have their children see that."

She notes that this is important, especially as tech struggles to find its place in the world. "I consider it to be a really amazing opportunity that way, to be living history and also to not even always know how it's going to unfold. Because that's part of what history is—you do your best. We're in the middle of tech backlash right now, which in some ways was not anticipated. But when we're working to ensure that, at least in my position, you simply do your best, that's part of what history is—it's people trying to get it right. And I think it's great to get to be part of that."

Regardless of our background or why we came to tech, once we are there we can begin to doubt ourselves. I kick off the next chapter with a personal story, like I will every chapter, of how I fully accepted who I was and that tech needed me to be me. This again is an important aspect of inclusion, that we can bring ourselves to our work and feel accepted.

2 | You Have to Be You

NEEDING ME TO BE ME

In 2013, I attended a training for women directors. A first of its kind at Google, it served as a combination of retreat and learning. Its goal was to help us figure out how to sustain high performance amid the unique challenges we faced. Why worry about this? An annual happiness survey had shown weak scores for women in senior positions, not necessarily surprising given the overall burnout rate in the industry[11] and even non-industry data indicating women face burnout at a higher rate than men.[12] The training was held at a nice resort, and I got a free massage. I felt like the winner from that survey!

The training couldn't have come at a better time. Despite my success, the truth is that I was struggling. At the time, I was a director of a growing team with two healthy children. But the question on my mind was still "Who should I be?" I looked around at my peers and saw many leaders, both male and female, who focused on business metrics or product definitions. You could throw a stone and hit someone who would debate a graph with you or pick apart data. Frankly, I was bored in those meetings. What I enjoyed was thinking about how to motivate people, how to build great teams, and how to grow employee happiness. Was I in the wrong place? Should I change my focus?

Throughout the training, as we sat together and spoke as leaders, I

---

11 Kelly, Mairead. 2019. "Tech Industry Burnout: Which Companies Have It Worst?" *Noodle*, December 10, 2019. https://www.noodle.com/articles/tech-industry-burnout-which-companies-have-it-worst.

12 Templeton, Kim and Carol Bernstein, Lois Nora, Helen Burstin, Constance Guille, Lorna Lynn, Margaret Schwarze, Neil Busis, Connie Newman. "Gender-based Differences in Burnout: Issues Faced By Women Physicians." *National Academy of Medicine*, May 30, 2019. https://nam.edu/gender-based-differences-in-burnout-issues-faced-by-women-physicians/.

heard what fellow women were struggling with, and their issues were interesting and exciting to me. At the core, so many of our problems *were* people problems, and that's what I really love. Another business or technical person wasn't going to be helpful at this moment; instead, it was my ability to listen and coach others that would be useful. This seemed especially true as Google was growing larger, and it increasingly faced problems of how to motivate and lead large groups. It was a real "aha" moment. I thought, "I have to be me, and Google needs me to be me." That set me on the course of truly embracing being a people-focused leader and letting that guide me both in the day-to-day and in strategic decisions, even personal ones like where I take my career. Knowing and embracing ourselves is critical but hard work, and this is covered more in chapter 8, "Our Magic Toolbox."

Why is belonging important? The term has only sprung up in a business context in the last five years, but belongingness—the instinctive human need to belong—has been studied in psychology for decades. In a paper in 1943, Abraham Maslow, a humanist psychologist, placed belonging, paired with love, at the center of his pyramid demonstrating the human hierarchy of motivations. This reflects our natural need for acceptance and relationships, which, if deferred, can lead to loneliness, anxiety, and depression.[13] I usually instinctively shy away from these terms, rolling my eyes at how we dress up simple concepts. I find myself adopting "belonging," though, because it does get to the root of why many of us don't flourish. How can we be successful if at the root of everything we do we don't think or feel that we belong; if every time we have an idea, we wonder if it's our place to speak up; if every time we disagree, we fear the downsides of sharing our thoughts? That adds up, and ultimately means we either don't act like ourselves or our jobs jail us without leveraging all our talents.

Let me give you that gift now. You have to be you, and the world needs you to be you. (I suppose unless you are an evil dictator, but I'll assume you aren't!) From what you focus on to your personal ambitions, let your

---

13 Cherry, Kendra. 2019. "The 5 Levels of Maslow's Hierarchy of Needs." *Verywell Mind*, December 03, 2019. https://www.verywellmind.com/what-is-maslows-hierarchy-of-needs-4136760.

inner voice drive you and not the definitions of the outside world. We need so much diversity in tech to build great products and services for the world. Don't let anyone convince you to be cookie cutter.

DESCRIBE YOURSELF

When I was interviewing women for this book, I asked a question almost entirely out of curiosity: How would you describe yourself in five descriptors? I was curious how women thought of themselves and what patterns or variation I would or wouldn't find. Do we label ourselves in traditionally female terms, emphasizing empathy and caretaking? Having worked in tech, do we now label ourselves in traditionally male or technical terms? I certainly would have labeled women I knew or worked with in many different ways; we are each unique, after all. But I'd never asked them directly, and this was my chance.

One pattern I found right away was that my question often made people uncomfortable, so I started disclaiming it and saying, "This question is supposed to be fun, but I find it makes people nervous." People started loosening up after that, but why does describing ourselves rattle us? Based on conversations with the women and my own guesses, it's really several problems.

First off, it's hard to think of ourselves as a classifiable object, whether we've done so recently or not. Many women asked whether I meant at home or work? They would struggle to define themselves at a point in time since they felt constant change. One example: "I feel like this is a hard question. Gosh. And in a specific work context or just a general 'who I am'? Because you know, sometimes you can show up to work differently than you are." Another woman said, "Here's what I'm going to be so bad at. I think my self-image changes so much depending on the situations that I'm in, that it's really hard for me to come up with a description that feels even remotely accurate in a way that's going to last longer than six minutes."

While we instinctively label others—our brain instantly sorts the world into compartments we easily understand—we may resist either positive or negative labels assigned to ourselves.[14] In particular, some women were hesitant to compliment themselves, wanting to avoid appearing braggy or egotistical. In those cases, they would search for words, hoping to find a humbler way to say what they wanted to say. One woman asked, "How does one describe themself? To me, that's always been 'you're boasting about yourself,' but that's not me."

I also felt some women would have been more comfortable if they could spend an hour on the exercise. To just list off some things about themselves that could go in a book seemed too flippant and random. Since I gave them this question in advance, some women took advantage and prepared ahead of time. I would often let them know that I was unlikely to describe them individually by their descriptors, and that they would be used more in aggregate for observations. Regardless, many women were more comfortable saying a term and then explaining it, to give context to what might otherwise be misinterpreted. This is a learned behavior. Research conducted at Yale by Corinne Moss-Racusin highlights the pressure women face to behave stereotypically, in this case to be humble. The study focused on mock job interviews where both men and women were asked questions that required self-promotion and talking about successes. Men were able to do this task more easily than women. Women tended to defer credit to a group or add negative disclaimers to achievements, e.g., "I struggled at first." While we'd like women to be able to take credit better, there is a downside if they do. According to Moss-Racusin's research, women who do are seen as bragging and thus less likable. This leads to real consequences, like being passed over for promotions and new roles or earning less money.[15]

It would be one thing if this only impacted how we described ourselves, but it also impacts our confidence, which leads to further repercussions on our careers and lives. Multiple studies have demonstrated this, including one from a group of researchers from Harvard Business School

---

14 The World Counts. n.d."What is Labeling Theory Psychology?" Accessed May 28, 2020. https://www.theworldcounts.com/happiness/what-is-labeling-theory-psychology.

15 Walter, Ekaterina. 2012. "Bragging Rights: Why Women Don't Talk Themselves Up and How to Do It Effectively." *Huffington Post,* October 22, 2012. https://www.huffpost.com/entry/bragging_b_2001545.

and the Wharton School at the University of Pennsylvania. The study asked a group of 900 participants to take a twenty-question test from the US Armed Services Vocational Aptitude Battery, which is used to determine who is qualified to enter the US Armed Forces. Having taken the test, the participants were then asked to estimate how many answers they believed they had answered correctly, and rate themselves on a scale from 0 to 100 for how they would score their overall performance. On average, women reported their performance as 15 points lower (46 out of 100) than the average man (61 out of 100)—despite the fact that the men and women performed equally well on the test. This was also true regardless of who would see the results; the women were more likely to rate themselves lower than the men regardless of exposure to a potential employer.[16] Despite the possible ramifications of doing so, it's critical that women see, embrace, and promote our successes, both so we truly feel how amazing we are, but also so others are forced to see it and recognize our talent. Breaking this cycle of how we've been "programmed" is key, and I'll cover this more later in the book.

Focusing on the descriptors themselves, I wasn't too surprised by what I heard. The most common terms were *curious* (fourteen uses) and *driven* (ten). Women also used related terms: *hardworking, passionate*, and *energetic* for example. In an industry with a focus on seeking solutions and driving hard toward success, people with these traits and preferences are going to be common. It also fits with the background of many of the women this book has introduced, who found their way into tech through curiosity and hard work. There was also a solid theme of positivity with a combined twenty-three uses, including *positive, happy, optimistic*, and *fun*. (All that marketing about tech isn't a lie—we really are a fun group!)

Of course many of the terms are exactly what we'd expect from a culture (predominantly US-raised) that raises women to be thoughtful, people-focused creatures. Common terms included *empathetic* (nine) and *caring* (nine) or related terms like *loyal, reliable, dedicated, flexible,*

---

16 Geall, Lauren. 2019. "Women are much more likely to downplay their achievements in the workplace, according to science." *Stylist*, November 2019. https://www.stylist.co.uk/life/careers/male-vs-female-employees-rate-performance-at-work-imposter-syndrome-study/310343.

compassionate, even-keeled, persistent, and *helpful*. We relate these terms to the feminine, and associate them with caretaking, giving, and patience.[17] The dark side of these terms is that they can also be underappreciated or considered weak, as reflected in a 2018 *Harvard Business Review* (*HBR*) article, "How We Describe Male and Female Job Applicants Differently,"[18] and multiple women did describe these terms as being a pro and a con.

But, as you'd expect from a culture that appreciates individuality, there were also terms we'd consider atypical for women, ones that are often attributed to the masculine identity: *ambitious* (seven) was the most common, with *direct, analytical, decisive*, and *fearless* showing up as related terms. Women sometimes used words associated with the less complimentary versions of these attributes, e.g., *stubborn, impatient, aggressive*, and various words that implied self-focus. The *HBR* article also notes this is common; women are often labeled by these terms when they flex the personality traits we associate more naturally with men.

What didn't show up that I expected? There were a few words that I was surprised weren't said at all, including *confident, assertive*, and *strong*, given that I would have described many of the women this way after hearing their stories. Sometimes variations showed up, like *aggressive*, but I expected the more positive versions to appear. I attribute this to the pains many women took to not appear as if they were bragging or complimenting themselves. They should, though; they are all awesome. I was, however, happy to hear mostly self-reflective comments versus damaging ones. When women brought up "negative" labels, there was more focus on how they could grow versus labeling themselves as categorically wrong. (This saved them all from a follow-up lecture on maintaining a growth versus fixed mindset.[19]) They were also reflective on why they

---

17 Pennycooke, Makeda. n.d. "Masculine vs. Feminine Leadership." Accessed May 31, 2020. https://makedapennycooke.com/masculine-vs-feminine-leadership/.
Voices of Youth. 2019. "Masculinity and femininity." August 20, 2019. https://www.voicesofyouth.org/blog/masculinity-and-femininity.

18 Hebl, Mikki and Christine L. Nittrouer, Abigail R. Corrington, Juan M. Madera. 2018. "How We Describe Male and Female Job Applicants Differently." *Harvard Business Review*, September 17, 2018. https://hbr.org/2018/09/how-we-describe-male-and-female-job-applicants-differently.

19 Mindset Works. n.d. "Decades of Scientific Research that Started a Growth Mindset Revolution." Accessed May 31, 2020. https://www.mindsetworks.com/science/.

thought they needed to improve, instead of blindly following feedback they'd heard without reflection.

It's worth noting that there was a very long tail of unique terms only individuals or a few women said. My favorites included *old soul, wry, a fake-it-til-you-make-it-er, bad at first dates, a fighter and fearless for those without voice,* and *a pessimist with hope.* I often related the most to these atypical descriptors because they spoke to what's unique in me, in each of us. Despite our commonalities, we are all special, and we each bring ourselves to the recipe of working in tech. There's no way to summarize all of that in several words.

Why did I go through all the work to look at these words and think about what they meant? In a way, it didn't really matter the exact words I heard, because tech needs everyone. We need people focused on caregiving *and* also those who are product- and business-focused. We need patient people *and* those who will push us to move faster. We need people who will never accept boundaries *and* those who will worry about what the far-reaching implications of our decisions mean. It's all these personalities that mix to lead to better business outcomes, environments that challenge us to work but are also fun, and finally that diverse and inclusive world we're all reaching toward.

Beyond the words, I've also selected some stories from the interviews that demonstrate the importance of being ourselves and bringing our unique backgrounds and talents to our work.

HISTORY'S IMPACT

How do our unique background and personality shape our journey? Amy describes herself as a nontechnical person in a technical world. As she ponders what she brings to the table, she reflects back on her father's background.

"I am, on a daily basis, grateful. My father was born on a farm in Taiwan, the second to last of nine. So poor and malnourished." She described how her dad got a PhD in mathematics and computer science in the US. "I could've been born on a farm with nothing easily. So the choices my parents made . . . it's unbelievable to come to a brand-new country, not knowing the language, not knowing anything, and building a life here." She connects this with why she didn't end up being more technical, and how she found her path in a technical world. "I was like, 'Oh, that's Dad's thing. He's trying to teach me Pascal; it's awful.' But the management piece is what I bring to the table. The ability to translate, ask the questions, lead, influence, deliver. That's something that not a lot of people are able to do, really focus on the delivery and bring a group of people to deliver a high-quality product on time."

Ginny Clarke is an executive recruiter and director of leadership staffing. As a black woman from another industry, she may never have joined tech if not for listening to her inner voice. After years of executive recruiting, she wrote a book to help guide people to take control of their career journey. At the time, she thought she wanted to be "the Suze Orman," a famous financial advisor, of the career space. As she grappled with what that meant, though, she realized that path wasn't meant for her. "I realized that some of it was because of some teachings from my mother, who only had good intentions. She was a broad-thinking woman from Tuskegee, Alabama. Her father had been the son of a slave. He got to Tuskegee and got his education there, put all six kids through there. Her brother was a Tuskegee Airman." With this background of pride, success, and hard work, Ginny's mom moved to Wisconsin, getting a master's and becoming a physical therapist. This all influenced how Ginny was raised, with humility being a key message. "You don't self-promote. You don't, and these were subtle messages."

While this initially seemed to hold her back, understanding her upbringing's impact also led her forward. "It was an interesting reflection on an influence and a belief system that someone else has for you that can become ingrained and hold you back if you're not really aware of it . . . I don't blame her for that, but I'm more aware of it now since I've done a lot of work to understand my own psyche, my own soul." Ultimately this personal journey opened her up to the possibility of tech, something she might have dismissed earlier. "This opportunity was presented to me, and I thought, 'Well, I don't care so much about tech per se, but I do care about diversity, and if this is considered one of the most nondiverse industries, if I can impact that at the leadership level, then it could have a trickledown effect and actually change the industry."

FINDING THE RIGHT ROLE

Similarly, our personality and preferences play a huge part in the roles we seek and where we feel successful. Christine Chau studied computer science, but she was always interested in how people think and what drives human behavior. "I really wanted to get into psychology, and the whole family was against it." After choosing to study architecture, she became interested in coding via writing scripts to survive in online games. She found the coding came to her intuitively, and that's when she moved to computer science officially.

When it came time to find a role, she interviewed at many places, looking for the right fit. "My mom actually wanted me to get into computer science from high school, and I refused. And the reason [. . .] I refused was [that] I didn't want to be sitting behind a computer all day again when I graduated." She decided to start as a systems analyst for Deloitte Management Consulting. "At the time software engineering was just coming to be. Software engineers were considered coders or programmers, as opposed to someone who thinks full life cycle. As a software engineer and being in management consulting, getting more involved in tech there, I got to be involved in the full cycle of software engineering." Ultimately her career led her to program management, a place where she could leverage her interest in the full cycle of product creation as well as how people get things done.

While never particularly technical, Hillary loves how complementary her relationships-focused role is to her engineering counterparts. "I think I catch people, especially like the engineers and technical people, a little off guard in a good way because I am so different in the way I approach what I do. It just works." Hillary describes how her focus is on relationships and connecting people versus the technical aspects of how products and business are run. She sees that her skills and perspective are different, and that's good. "My job is about relationships, and if I can relate to an engineer coming from a completely different perspective and asking questions that they then feel they can share without having to talk about the technical aspect of it, then I think it's refreshing for them to have that conversation with me and make it about a relationship."

INDEPENDENCE

Ching's journey contains critical moments where she decided to make a change, regardless of what others thought. At college she was studying microbiology, which "turned me into a total germaphobe." She took one business course her senior year, Entrepreneurship. "It grabbed me. It was a really interesting experience to be in a room where it was all dudes, and the group of us that decided to form are four women." In the class you came up with a business idea, and the final project was to present it to VCs. "We go through this experience, and it totally changes everything for me. I was ready to go off and become a registered dietician, like I got into grad school for all of that—on my way. Then I had to have a really difficult conversation with my parents that basically went into, 'yep, so I'm not going to do this. I'm going to go to a really small boutique management consulting company for a few years, and then I'll figure out if I want to come back and get my RD.' So I went off and did that and I never looked back."

Her career took her to the pharmaceuticals industry in New York, where she worked for a decade. But she started to lose the joy of the work there. "I was just showing up to show up, and I was doing a great job and had gotten lots of promotions and all this and added responsibility. But none of those things actually made me feel any better. And so I decided, I'm going to quit." Ching quit and took a break to travel for a year. When it was time to return to work, she wanted something new. "I don't want to keep doing the things that I've done. And so I thought, what's a place I'd really like to learn more about and be a part of? And Google popped into my head. It was totally outlandish. I talked about it with a couple of friends and family, and they were like, 'You don't know anything about tech.' And I was like, 'Exactly, but I know nothing, and I want to learn about it.'"

With limited background and a desire to be on the engineering side of the business and avoid jobs requiring an intense travel schedule, Ching found an administrative assistant role. "I remember when I applied for the job, the recruiters were like, 'No, you're using this as a stepping stone to get into Google, and you'll get in. Six months later, you're going to get a different job.' And I said, 'No. This is what I want to do.'"

Ching identified one characteristic that has been key in her life: being stubborn. While joining Google and becoming an admin was something she wanted to do, "It's also caused a great deal of grief in my life." Despite being an interesting career pivot, becoming an admin had implications. "Everybody was like, 'You've done all this. Nobody goes from what you have done to becoming an admin.' And I was like, 'Why not? There are interesting things to learn there too, right?' But there's this expectation that people are maybe not as intelligent in that bucket." Ching believed in living the way she wanted to. "And if a lot of other people want to put [out] their expectations or their judgment, they can leave it at my feet, and I will continue doing what I need to do."

Ching proceeded to learn and grow on her own terms. She started her new role and found a flexible environment where "you could pretty much contribute in a way that you wanted to contribute, as long as you let the folks know what you were interested in." Leveraging this, she picked up work across program management, analytics, and strategy. "I do feel like I was able to develop a closer partnership with a lot of the senior leads because they were able to almost use me in a different way." Ching parlayed this experience into a role as chief of staff in her VP's growing team.

EMBRACED AND EMBRACING

For Jill, part of her identity as LGBT has contributed to how she defines herself. Part of why she likes tech and has felt supported is that it's a very "queer-positive place," which she's not sure she would have found in more traditional industries. "As the LGBT person, I think that was also a piece. There's gender, but there's also gender expression. I do feel like that, as a general matter and certainly [in] my direct lived experience, it really [is] the best of tech. It truly is a meritocracy." As a disclaimer, Jill does note that she's not a salesperson on her own, selling into a hard territory. She knows experiences may vary, but overall she feels "this [is] going to be the legacy of Silicon Valley, specifically in California especially: the embrace of any and all who have good ideas." She also acknowledges that she does have privileges as a white person that people of other races do not. "Even though tech gives us much more latitude to lead with our contributions, we still have so far to go in terms of equity and inclusion and addressing racial inequality."

Sandra Montalvo Leppek initially felt uncomfortable as a Latina woman in tech, especially starting her first role in Seattle far from the Southeast where she grew up. "I think more and more in today's tech scene, we're sort of in the aftermath of everything evolving, like diversity and inclusion issues. I think it's much more about not letting things change you. You can absolutely be yourself and be whoever it is that you want to be in these companies. There's space for you." She notes that this isn't always how it appears at first. "When I first started, there were so many Asian American women in the office where I sit regularly. I was super self-conscious about my hair because it's big and frizzy, and I'm big and curvy and loud." She felt imperfect among a sea of women who seemed to lack wrinkles and frizz. Feeling like she stood out, "my initial instinct is to hide myself. I will straighten my hair and blend." But then she thought, "That's not why they hired you. They hired you because you are you. So stop that. Get those thoughts out of your head." She's discovered that embracing who you are is key, especially because you may be the first of others to come. "You just maybe are the first one to get to where you are." Sandra's point is key: you need to be you in order to encourage others to be themselves as well.

Tequila Brinson also discovered this in her career journey. She has been able to leverage a company program for coaching; "it's for people of color to harness and dig down deep to places that are areas of our lives where we have not gone, that are kind of broken, and pull that out of us." Being coached has changed her life both as an employee and as a black woman, and now she wants to be a coach as well. "It changed my life because it allowed me to really show up as this quiet girl, but be transparent and unapologetic for who I truly am, where I come from, and not be ashamed of the color of my skin or the woman that I am or where my parents come from and how they raised me. I am no longer afraid of that. I mean, it is who I am, and I can't change that." Now she wants to channel that into helping others. "I'm a coach to coach people through the things that they are struggling with, the places that they're stuck, the places [where] they're embarrassed, the places that they don't want to be vulnerable in, the places [where] they feel shame. I want to give them the tools to be able to get to that place."

These three women, among many, remind me how important it is that we be who we are, and that we embrace our differences instead of assimilating or ignoring what makes us unique.

REDISCOVERY

But sometimes we forget that or we lose our way.

Sharon Park is now president of her own ad agency. Her journey in tech has been one of rediscovering herself. While working in tech, she felt like she'd reached a ceiling. She wanted to lead a big sales team and left to do so, but she returned to her previous company because she enjoyed working with the people there. After a reorganization of her team, "I looked for another position within the sales team and couldn't find meaningful work—which seems illogical given our growth rate and hiring needs." She also discovered she was paid considerably less than a male colleague who had started after her, also entry level. She was triggered into self-doubt and anger; how had the spread become so large over time?

Ultimately, she was pushed into the worthy but difficult exercise of scrutinizing what success meant to her. "I have been mulling over the meaning of success. Is it odd that when I was [there], I felt cloaked in failure, but now I own the tiniest company and I feel alive? Every bite of food (that I now have to cook for myself) tastes better. The sky is definitely bluer." She's learning to ignore the call of outer success: "And this is a very loud concept in society. Money, power, fame, influence, status. This definition of success is everywhere around us." She decided to instead focus on her definition of inner success: "Time with my family, feeling like my peak talents are at work, feeling actualized, time to care for people in my community who need me, and time to be creative and explore physically, emotionally, intellectually, and spiritually. At the core of inner success is family and my partner. They are my source of my energy to do work." Finding her motivation has been the secret to recapturing her excitement and joy.

I love reading these stories of how women have found their way and embraced who they are, especially as they persevere through doubt or struggle. It reinforces the part of me that needs to remember that who I am is enough and exactly who I should be. Our journey continues in the next chapter as I tackle the age-old question we all face, "What do you want to be when you grow up?"

3 | We Have Different Ambitions

REWIND

I grew up without much money. The way I describe it is, I never went hungry, but there was more than once when we didn't have enough money to pay for all the groceries and had to put items back. I've never had someone not nod when I describe it that way, and I think it's because the embarrassment of not having enough money in a grocery checkout line is so public and visceral, even if you haven't experienced it. My favorite part is everyone waiting behind you, judging your priorities. That is not a good time to keep the ice cream.

Once when I was a kid, maybe seven years old, my mom and I were shopping in a store similar to Target. I found a My Little Pony toy that said it was $1.00 on the tag. I knew instantly this was too good to be true, but I marched it over to my mom and begged. Normally good toys were out of range for everyday purchases, but maybe this once we could make an exception. At $1.00 my mom didn't say no, but I saw hesitation in her eyes. She knew what was coming. When we got to the counter, the toy rang up as $11.00. It was mismarked, and we couldn't afford to buy it. I've told this story a few times as an adult, and the reaction of others is always that the store should have honored the price tag. And yes, it totally should have. But what I know is we didn't really ask. We were used to disappointment, and we knew we wouldn't get away with it. Luck didn't bend that way for us, and I had already accepted that as a child.

When I was in sixth grade, I applied for and won a scholarship to the school I attended. For years my parents had paid full tuition to a private school, so I could get a better education than my local public school. Here finally was a way I could help our family, and after an essay and interview process, I was the winner. It was a rare and delicious moment of success. My dad took me out to dinner, and I remember him kvelling as dads do. But then he said, "Things like this don't happen to people like us." He didn't mean it badly; he was honestly marveling that life could be that good.

I can't rewrite my history; it's part of me. How it connects to my career feedback over the years may surprise you, though.

FAST-FORWARD

At various moments in my Google career, I've been told to be more ambitious. This feedback was intended as a vote of confidence. They saw me as a skilled leader who could move up the ladder or even start her own company. Sky's the limit! Anything is possible!

Google has a performance review process every year where I receive feedback from peers. From the reviews I've gotten over the years, I can see a pattern:

"It's hard to do the same job for as long as Alana has and remain optimistic about future personal growth. Alana has to identify what will give her the greatest personal satisfaction and demand a career path that supports her growth."

"Alana simply doesn't realize how veteran and senior she is at Google. There is so much she could do if she realized how skilled she is and recognize[d] that having the many abilities outlined above gives her the ability to truly play a leading role."

"I think Alana could be more ambitious. She is one of those rare people who is far more capable than she even knows. I think she should think bigger and shoot higher in her next role."

I appreciated their future vision of me, but I didn't see it. It's like when someone tries to tell you about a country you've never visited. You can marvel at what they saw, but you can't truly imagine what they're talking about. Even photos are pretenders; there's nothing like the real thing. Ambition is my faraway land.

WHAT IS THIS THING?

Possibility is a luxury—you need to be able to afford it. Some of us are raised with ambition as an everyday thing; we see it in our family and friends. Others have to fake it until we make it. We're climbing that ladder rung by rung, and we frankly don't care where the ladder goes as long as it's up.

I was successful because I was a hard worker and I made good decisions, and that helped luck roll in and do its thing. But once I had climbed into a better life, I had no answer to the question of what's next except to stay there. I'm the opposite of what people talk about in the business world. What if you're just happy to be here at all? Just thankful you don't have to put things back from your grocery cart? Grateful you can stay in a nice hotel when you travel? What drives you then?

It turns out fear drives me. I love security. I often joke about how I am essentially Scrooge McDuck; I would love a room filled with money to swim through. I never ever want to feel the stress of credit card debt or school tuition again. My husband can tell you I have a hair trigger about money. Paired with that, I am interested in a different life. I watched fellow coworkers become COOs or CEOS, and it wasn't my ambition. I felt jealous of the drive, but not of the jobs. I like small teams, creative problem-solving, and building expertise. While I might want my own small company one day, I'm on a different path than them. Sometimes I would beat myself up about staying at Google and still being a director versus in the C-suite, and then I would have to say to myself, "Wait, you don't even want that job." And that's okay.

In fact, it's great. I've had four different jobs at Google that were incredibly interesting, and I've been rewarded both financially and via other recognition, including my personal satisfaction. While my job has sometimes required long hours, it's not all-consuming like some roles can be. I've been able to prioritize my wonderful family, and people compliment me for my "perspective" and "balance."

I'm not jealous of other jobs, *but* I am jealous of how effortless it seems for others to dream and want bigger things. Am I missing out? Am I alone?

WHERE LIFE TAKES US

In speaking with women for this book, I saw that I was among company. While some women aspired for a C-level position or the next rung on the ladder, often women were looking for titleless objectives like learning, being excited about work, developing others' careers, or having time for family. While we certainly do need leadership ranks to be more diverse, it was interesting to see how many women's career ambitions focused on other areas.

Data backs up this observation. In a 2015 Statista survey, nearly 40 percent of women responding either disagreed or neither agreed nor disagreed with the statement "you like to set career goals for yourself."[20] In another study published by PNAS in 2015, when asked to share goals, women listed more life goals, e.g., being in a committed relationship or being organized; however, a smaller percentage of those listed by women related to achieving power. Women also viewed high-level positions and promotions as equally attainable but less desirable than men did.[21] And certainly career goals change over time, with job security, ability to do what you love, or work/life balance having varying importance depending on our life situation.[22]

This presents a catch-22. We want to retain women in tech, especially in leadership positions, not only to support other women in the reporting chain but also because studies demonstrate better results and increased innovation with women-led teams.[23] But what if life takes them elsewhere? And why is life taking them elsewhere? While motherhood is often cited

---

20 Statista. 2015. "Women who like to set career goals for themselves in the U.S. in 2014." Published Jaunuary 31, 2015. https://www.statista.com/statistics/607857/women-who-agree-with-the-statement-that-they-like-to-set-career-goals-for-themselves-in-the-us/.

21 Gino, Francesca, Caroline Ashley Wilmuth, and Alison Wood Brooks. "Compared to men, women view professional advancement as equally attainable, but less desirable." Proceedings of the National Academy of Sciences of the United States of America (PNAS) 112.40 (2015): 12354-59. https://gap.hks.harvard.edu/compared-men-women-view-professional-advancement-equally-attainable-less-desirable

22 Roberts, Jeff. 2014. "Survey: 'Most Important' Career Goals Change with Age." Rasmussen College, July 31, 2014. https://www.rasmussen.edu/student-experience/college-life/most-important-career-goals-change-with-age.

23 Kemp, Leanne. 2020. "Having women in leadership roles is more important than ever, here's why." World Economic Forum, March 3, 2020. https://www.weforum.org/agenda/2020/03/more-women-in-leadership-shouldnt-matter-but-it-really-does/

as an obstacle, data reflects extreme work pressure; hostile, macho culture; and compensation as issues.[24] And what about the women who do want to drive for the top? The way there is often unclear and fraught with obstacles that can lead us to doubt ourselves, like being held to higher standards or having fewer connections or sponsors.[25] When women do leave tech, they don't necessarily leave the workforce as sometimes assumed, instead pursuing other options including work in other fields and self-employment.[26] We will revisit how women stay or leave in chapter 11: "The Art of Staying."

I should note that I speak with men about their careers as well, and much of the above also holds true for them. The business world can often seem obsessed with competition and the drive to the top positions, but many people look for a more nuanced approach to their careers. In recent years, in particular in conversations with my peer group of forty-somethings, I've noticed how many of us have completed our initial ascent in our careers and are now searching for what lies beyond. Much as David Brooks discusses in *The Second Mountain: The Quest for a Moral Life*, many of us complete a first career trajectory to later seek another path. As Brooks says, "Eventually there's no escaping the big questions. What's my best life? What do I believe in? Where do I belong?"

PERSONAL GROWTH

I asked several questions that helped me understand where women were in their careers and what they were looking for. One question was direct: "What is your career goal now?"

Other questions often gave me color or additional insight, e.g., "What's

---

24 Hewlett, Sylvia Ann, and Carolyn Buck Luce, Lisa J. Servon, Laura Sherbin, Peggy Shiller, Eytan Sosnovich, Karen Sumberg. 2008. "The Athena Factor: Reversing the Brain Drain in Science, Engineering, and Technology." *Harvard Business Review*, May 22, 2008. https://www.talentinnovation.org/publication.cfm?publication=1100.

25 2015 "Women and Leadership: Public Says Women are Equally Qualified, but Barriers Persist." Pew Research Center, January 14, 2015: 31 - 39. https://www.pewresearch.org/wp-content/uploads/sites/3/2015/01/2015-01-14_women-and-leadership.pdf.

26 Bailey, Kasee. 2020. "The State of Women in Tech 2020." Dreamhost, March 6, 2020. https://www.dreamhost.com/blog/state-of-women-in-tech/.

a key challenge you have faced in your career?" or "Do you feel like you're surviving or thriving in your current position? Why?" Far more common than a specific answer when I asked women these questions was a feeling—women want to feel like they are growing and learning. This career goal is often broad and open to possibility. The desire to remain passionate about the work is a strong motivator, with learning being one of the key contributors to remain enthused. More than financial goals, these individuals want work to stimulate their brains and lives.

Adrienne didn't know what she wanted to be when she grew up. "My mom was a big influencer to steer me toward math and science, which ultimately led to engineering and college." Adrienne is in digital marketing, and her career goal now is to "learn and grow personally. It's not about hitting a certain position, a certain salary or level. It's more about 'how do I continue to be excited and passionate about the work I do and actually personally grow?'" This wasn't always her answer. She says that if I'd asked in her twenties, she would have wanted to be a leader, something specific like having a team of fifty people. But now she's already led a team, and her goals are less concrete. "I kind of did that, and what I really want to do is learn some new things and go in a different direction, but I wouldn't have thought that a few years ago."

Marily has arrived in a similar place as Adrienne, although they focus on different fields within tech. While pursuing her PhD, Marily did internships at a different company every summer. "I worked as a data science, software engineer, and web developer intern, and got to experience three very different tech companies, which gave me a well-rounded understanding of how large corporations work." She knew that she wanted to join one of those companies full-time and had two mentors who helped her understand the companies and the roles and prep for interviews. "I had never heard of product management before, but when I found out that it involved engineering, problem-solving, [and] having a vision for something new while also setting the strategic direction for the solutions I came up with, I knew that this was what I wanted to do." Her career goal now is pursuing this road of personal growth and inspiration. "It is very important for me to keep growing as a computer scientist, to keep being challenged, to keep learning things, and to keep making mistakes! I live a life surrounded by tech, but I want to live a life and career surrounded by innovation, as this is what challenges me: making users' lives better and easier by solving problems they don't know they have. As long as I can innovate by using new and current technologies, I get inspired, and that's my goal."

MOVING UP

Many women also want to keep progressing up the ladder, but it's rarely at any cost or without other combinations of goals. For example, Camie would like to get to VP as far as promotion, but she doesn't want to sacrifice having fun and enjoying herself in pursuit of that goal. "I enjoy building great products and building great teams." She still wants to feel the excitement of coming into work every day.

Kris has a specific goal before she retires: "to be the first VP woman without a degree" at her employer, which she's fairly certain has never happened before. She is particularly focused on achieving that at her current company. She finds it a good cultural fit, and a "personal growth–focused company that I've never encountered before and I don't particularly want to leave."

Natalia is ten years into her career and is focused on making the most use of her potential. "I'm definitely seeing how far I can go in my current world and my current role. I'd love to get to the next level internally." At the same time, she's aware she could potentially leave her current employer if there were better opportunities outside. "I'm aware of other compelling tech companies and opportunities they offer and their expanding presence in San Francisco. My goal is to make good business decisions and use my time and resources and skill set wisely, to drive the most impact per unit of my time."

HAVING IMPACT

Closely correlated with growing themselves, many women also want to see and feel their impact. Given many women joined tech to change the world, this is often a natural extension of their original motivations.

Whether they are now shifting focus to embrace a new area where they can have an impact or leverage their seniority to drive their ideas forward, they want to be the owner and driver of important work.

Jill envisions a CEO role in her future because she wants to grow and have accountability for others. "What I love most about what I do now is I really love helping people become their best version of themselves or whatever it is they want to be. That is something that's really important to me. I think next for me it would be the CEO." She was previously a CEO for a small company earlier in her career, but now she can leverage her broader and deeper growth amid different-sized teams and initiatives. In particular, Jill wants the responsibility of driving an entire effort and team. "The thing that is appealing to me about being a CEO is really that it's this ultimate accountability for delivery. And that to me is really interesting, being able to set that tone culturally for an organization and effort."

Caragh likes the idea of moving toward a director role so she can own a business area. She wants to be responsible for "everything from the P&L to the strategy, and I could own the bigger picture." To some degree, she's flexible in what area that falls, and she's not hung up on a promotion per se. The larger story is "I'm driven and I want to continue to do well in my role and find new opportunities where there's new problems to solve or things that I can make an impact in."

On the opposite end, Annie gave up managing in her most recent role. Moving into an individual contributor role has been "kind of a shock." At the same time, "it's also really exciting because that's where I felt like I could really make the most impact right now." She doesn't necessarily have long-term goals. "I do think it's good to have goals, but I see [those] more as, what do I want to learn? What do I want to achieve? What do I want to help build and support?" She derives a lot of joy from supporting other teams because "then you get to multiply your successes times a hundred people when you see them all succeed." She says firmly that she doesn't need to be CEO. "That's so stressful." Her goal is to feel challenged and to feel like she's making an impact.

Sarah focuses on impact, whether for herself, at work, at home, or in her community. "I want to make a big impact *while* prioritizing my values, family, and life. And to earn an income that allows my family to not have to worry about money. I grew up in a family that had money issues and felt the strain of that as a child. My biggest goal is to ensure my kids never feel that pressure and they're able to reach their own potential without those barriers at home. I don't have a title or level I'm striving for, but continuous growth is critical because I thrive and get energy from growth and change. I also want to have side hustles that impact my local community."

HELPING OTHERS

Ashley spoke about another common element: wanting to aid others. When we met, she was performing a job search after a break traveling. "I'm talking to some recruiters to get a feel for what kind of companies they are and what kind of companies I could see myself working at. I think I'm trying as best as I can to hopefully replicate what happened at my first job and enter into a smallish company." Ashley likes that environment because she can immediately have an impact. When she started at LendingClub, work was highly manual, with spreadsheets and notes. "And I took all that and automated everything." She likes the feeling of impacting people, of making their lives easier. "It was a good feeling."

Melanie is looking to have that type of impact. Despite going through a period of career change where things are uncertain for herself, she really wants "to find a place or a role where I can help create an environment where people are really happy about what they're doing, they have a good work/life balance, and they're delivering value and making the company really successful while enjoying it." She notes the importance of the environment we create for others and wanting to have that "influence at scale."

Karen Wickre worked in communications at Google and Twitter, and wrote a book about networking for introverts. She is now sixty-eight years old, and she can see the arc of her career over the last thirty years and even now feels like her career is evolving to leverage her skills to help others. "I'm doing some consulting and thinking about things I really like to pursue where I'm still an explainer in a way and a conduit between people and seeing around the corners." She gives an example of meeting with a young company that needs help finding the right communications agencies and being the liaison so the work gets done correctly. Karen doesn't necessarily know what to call that role yet, but "it's drawing on my connector skills, in a way."

HAVING A LIFE

It's not always about work. Some women were clear: first and foremost, they work to live. Many women saw their goals change over time as their life changed or as they achieved certain milestones. Most of these changes were natural, based on the phases of their lives versus feeling as if they'd been forced on them by lack of opportunity or restrictions. Like Diane, who said, "My career goal? I think it's to be happy with my job and my life. I don't really have a set goal of get to this level or that level or become CEO of a start-up. It is more to do something that I enjoy doing and allows me to have a life."

Bethanie's focus is having a career where she can feel supported in having both a work life and a family life. "In terms of immediate career goals, I think maintaining this balance is really important to me. I hate to use that term, but it really is ebbs and flows." She is thankful for the leadership role she has in business development and partnerships. "I feel so blessed that I'm able to have a career that is maybe at times not the most fulfilling, but allows me to support my family and also allows me to be home for dinner every night and be really present in my kids' lives when I want to be. I think when I look at counterparts of mine or friends of mine that have demanding jobs or similar earning power, their personal lives suffer." She feels this acutely now as her children are growing older. With her twelve-year-old, "like, I don't know when he's going to be ready to talk to me about things he needs to talk about. I want to be there at that moment." She's also focused on the near term, admitting she doesn't have a three- to five-year plan; rather, "I just want to get through the summer. I want to get the kids back into school and try to think about things a little bit more one day at a time." That said, she never has been the type of person to have a long-term plan or even be a businessperson. "I think my ambition in my career has surprised everyone in my life, including me . . . I was not Ivy League; I didn't have that type of vision for myself. [My career is] kind of a happy accident, and I'm excited to see what other kinds of happy accidents evolve." For now, Bethanie has the flexibility she needs, and that's contributing to her overall career and personal success.

Reese thinks of her career trajectory similarly, saying it's less about goals and "rather, how to have both impact while leading a balanced life." She admits she hasn't figured that out, but "if you sort of asked me professionally what I care about, I'll often use the word *impact,* because I try to look for things that I can do that move a needle and have looked for things [at her company] that moved the needle on societal issues." More and more she is asking herself, "But who am I to my family and who am I when I'm at home?" Answering that question, while still having impact at work, has now become one of her life goals.

Kathleen is earlier in her career journey and isn't sure what lies ahead, but she's thinking about family and her lifestyle. "I'm not great with long-term career goals. I feel like in the past I've set them and I ended up changing the mark, so I keep my career goals pretty short, as long as I'm learning and engaged. . . . And I feel like until now, I've not been prioritizing home life. Not that I've really had to, but I want to make sure that I can do it. So I think my current career goal would be to keep having the impact I'm having but also do it more effectively or efficiently."

Leanna jokes she wants to retire, but it's more that she may want to explore different lives and careers. "I think that my career goal now is to achieve a certain level of financial comfort and then retire into an interest. And I think that that's a pipe dream, but I'm still a little bit of a tourist, and I still every six months want to open my hospice for dogs and my bakery somewhere in Montana." From a more practical viewpoint, she thinks she could freelance and work remotely anywhere with steady Wi-Fi and is looking forward toward that measure of success.

Connected to life goals, many women are measuring career success via how they can take care of themselves at the same time they have a job. Kristen notes that "walking to work has been a career goal for me for a long time, and I do actually walk to work. It's less than ten minutes from my house." This was a big change after commuting in the San Francisco Bay Area. With that done, Kristen feels like her goal now is to "have a big impact and bring a lot of people along on the journey." She's paying attention to how she helps others build their careers and gives them opportunities via "mentoring and coaching, and trying to open doors for people."

RETURNING TO OUR ROOTS

When I have a career conversation with my team members, the first question I ask is about what they studied or liked studying in school. When we look back, we tap into our original motivations and dreams before life may have taken us down other roads. It's useful to take a peek and see whether there are still seeds of ambition there to grow. In my case, I wanted to be a writer and an actress when I was little, a storyteller. Revisiting those goals has centered me; it reminded me of who I really am versus what Silicon Valley says I am. And I've refocused on writing as a result. Other women are similarly looking back to look forward.

Camille's latest career move has brought her into the real estate and facilities space. Her hometown is Savannah, Georgia, and Camille has a potential interest in taking advantage of its tourism popularity for a future business venture (years ahead and just an idea for right now). She's leveraging her current role to learn about property management for a potential career change later. "You know, end of career. It's a great place to retire, anyway, like the next career after I've learned more about all this property management, facilities, and the bones of buildings."

Alex studied computer science at Northeastern University in Boston. She wanted to be an inventor at first, and had a great interest in mathematical puzzles. "I took my first computer science class during my senior year of high school. I then majored in CS at university. I started my work at SmarterTravel (a TripAdvisor subsidiary) as a web developer co-op for my first internship." For now she wants to invest in her technical skills and become an expert. "After having a solid technical foundation and becoming more financially stable, I'd like to start a nonprofit coding camp for minorities and lower-income families. I would like for them to have an opportunity to earn well and diversify the tech industry."

EXPLORATION

While some of us are revisiting our past for inspiration, others are striking out in the search for new ideas and places. Ashley is a great example. She was reaching a point at work where she was feeling too comfortable. After the rapid growth she experienced at her start-up, she felt embarrassed to be thinking of leaving. "I was like, *Don't be spoiled; you have it good*." But when her boss, whom she really looked up to as a great teacher and advisor, quit, she decided to leave too.

Ever since college, when a friend had traveled to Asia for a few months on a one-way ticket, she'd had that idea stuck in her head. "I always wanted to do it, but I was too scared and never thought I would be that type of person to do that." She had a partner in crime, though, as her boyfriend was also ready to quit his job and leave. They booked one-way tickets with a vague idea of which countries they would travel to, and two months turned into five months. They ultimately visited nine countries, including Thailand and Vietnam, where they stayed a month each.

At the time I interviewed her, she had recently returned to the United States. "We've been apartment hunting, and slowly I'm facing reality and realizing I need to get a job again." She admitted she was "super scared" about finding her next role, and she again found herself in the role of explorer. "I feel like your first job out of college, you were like, 'I hope I can get a job.' And then your second job, you realize, 'I could have options.' It's not just about a company willing to take me. It's also about me choosing the right company."

Cathy is a freelancer focused on user experience and design. She stopped working at traditional companies after giving birth to her daughter, and she loves doing user experience (UX). "UX to me is not just a job, but a way of thinking and seeing the world and being able to use those abilities in a way to enhance and improve whatever systems or anything that you're building or touching." At the same time, she's very interested in food, and she's looking for ways to bring her interests together as she works her freelance jobs.

HITTING A WALL

Some women have been working hard, driving toward the top, and then they realize it's unclear what's next. Having to define and clarify their path is the next challenge.

I met Amy at what I call "summer nerd camp" when we were seventeen years old, and I've watched as she pursued her career in tech. While I was writing this book, she shared snippets of awesome career milestones with me, like presenting a talk called "The Value of Mentoring" at United Way's worldwide headquarters. At work, she learned she'd be overseeing the majority of the East Coast region in her technical operations leadership role. From my vantage point, she seemed to be headed straight to the top and marketing herself like she wanted to get there.

As we spoke, however, I picked up on insecurity about her lack of technical background, i.e., coding. I asked her about it. "I do. I feel like I've hit a wall. I don't think that I'll ever be a CTO." She's leaning toward a chief of operations (COO) role where she could use her broader range of skills. While she's not formally educated in technical subjects, Amy has an MBA. She emphasizes her ability to prioritize and make sense of work, to shield her team from chaos, and to figure out what needs to be delivered. "Because there's eight hundred million things being asked of my team, but does it actually make sense from a business point of view?" At the same time, she's decided she's not that interested in the CEO role anymore. Going into business school, she was, but now that she has visibility into what they do, "I'm like, 'Ah, I'm not really sure.' There's no one really above me right now where I'm like, 'Oh, I'd like to do that.'"

She also has a strong focus on culture, inclusion, belonging, and diversity. "There seems to be a lot of opportunity there with people, companies hiring disruptors, and chief D&I officers. So is that something I can sharpen and style? It would be valuable, and that's the work I do . . . how you include everyone, how you bring everyone to the table, how you allow people to thrive in an organization." She already runs two successful diversity and inclusion programs at her company. She's taking this into account as she contemplates possible paths forward in her career.

Jessica T. is feeling the frustration of diverted dreams. As she speaks about her career goal, she starts to tear up. "I think ultimately my goal is to be a CMO, but I'm not so excited about it when I look at CMOs, especially at public companies. I don't necessarily want that for my family. I don't want to be on a plane all the time." While seeking to balance her family life with her work life, she hasn't been able to find the role she seeks. "I want to be able to practice my craft, not just sit in meetings looking at PowerPoint decks. I think probably a CMO at a start-up would be ideal. The reason I get weepy is because I definitely feel frustrated about where I am."

Jessica explains that she thinks she is getting farther away from her goal, not closer. This is partially due to choices she's made in her career, but she predominantly feels that her experience isn't recognized. "Or maybe I haven't figured out how to channel it, in a way. So, I don't think I'm going to achieve that goal." As we talk more, Jessica also talks about how she feels like her age is playing against her because she's seeing fewer opportunities than when she was in her thirties. "It was a really different experience from being 31 and coming out and having a lot of offers at that age. I think there's something about being a freshly minted MBA; it might also just be being a woman in your thirties versus in your forties."

Frances feels a very strong pull from the two major roles in her life: mom and senior leader. "I have always been a 'driver driver' in my career. I went and got my MBA. I've always been a climber, and now I'm struggling with wanting to be able to take on more, wanting to learn more, wanting to make more money, quite honestly. Because I like it, and it's expensive in the city." At the same time she's trying to balance that with wanting to be around for her children. "When my girls are going through stuff, I want to physically be there to hear them." She's cut her travel over her last three jobs from 80 percent to close to 20 percent, but that has come at the price of her scope.

She's not giving up on further career goals, but she is wondering about when in the future she can push for them. "I've thought about eventually becoming a COO. I feel like I've done a lot of the work that goes into that, but I don't know when that will be possible. I think maybe when both of my girls are in high school, and I'm more comfortable with the fact that we have put a foundation of confidence, of self-esteem, of street smarts [in them], that I'll feel more comfortable with working longer hours again and working on the weekends, because that's what it requires."

FIGURING IT OUT

Finally, there are those of us who are figuring out our careers as we go along, and that's also okay. Careers are long, and sometimes we will spend time testing roles out, learning what we like or don't like, and experimenting in new areas.

Sherry is on that trip now. "I had a career chat last week, and I pulled out the *Designing Your Life* book . . . I was panicking about how to fill an hour's worth of a career chat [with my manager]." For the initial years of her career, she was focused on international business and expansion after three years living abroad. She discovered that "internationalization has a component of things I love, but it's not the only thing I love. It was helpful to realize that my focus doesn't necessarily need to be that narrow."

She discovered this while working as a product manager on phones and finding it fulfilling, even though much of the work was not international. "I think a lot of it has to do with translating technology for people where technology wasn't necessarily created with them in mind." She saw that helping people who weren't familiar with technology at all—whether due to geography, language, fear, or lack of understanding—motivated her. She sees her main career goal now as helping to bridge that gap. "It's a very long-winded answer to 'What is my career goal?' But I think that's where I see myself continuing to grow."

Kerry has already been a director previously, and she's now a senior manager at her current company. Her goal isn't to get back to that position, though. "I'm not looking for this next promotion, not dying to be a director or anything. I think I love what I do right now. For the next few years, given the space that I'm working in and how much it is evolving, I want to see where it takes me and continue helping other people grow within this space."

Similarly, Michelle wants to keep moving and exploring without a definitive destination. "It's hard to explain out loud. Let me try. I vaguely want to get promoted another few times, but it is not for the goal of having a higher job title. It's for the goal of having to learn skills at higher and higher levels of abstraction." She gives examples of making the jump to her first management role and then another leap when she started managing managers. She enjoyed the growth that happened with getting over those hurdles, and that's why she stays at a large company. "It's a good place to try learning those things, because if you go to a small company, you hit a ceiling at some point just by the number of people."

We've completed part 1, called "You Belong in Tech." As I regaled you with examples of how different women have come and found their way in tech, my main goal was to build your confidence that *you* belong in tech, whether it's because you're considering joining a technology company or because you already have. I believe there's a place for you, and I'm not-so-secretly pushing that agenda.

At the same time, I want to be balanced and real. There are challenges of being a woman in tech, and we shouldn't shy away from having those discussions. Part 2, "The Pros and Cons of Being Rare," will dig into those stories next.

PART TWO

THE PROS AND CONS OF BEING RARE

Ching
I think that's what happened with the generation of women who are rising right now: we have created a different level of awareness where it's not just about us, it's about the future generations and having to change the game for them.

Alana
I often think of it as a win-win. Like, I'm going in and I'm being a pain in the ass around my salary. Because if I'm not a pain in the ass around my salary, I'm letting down other people in my situation, and not just women but people like me that will take big job leaps and take big risks and do things that other people wouldn't have done. If you can't recognize that as valuable, then we're hurting them too. And I've spoken to both men and women who feel hurt by that.

Ching
In a lot of ways, you're trailblazing.

Alana
I'm trailblazing, everyone! Let's put that in the book.

Ching
But it's true. If you don't create a path, others may not have as much of an opportunity to walk it.

4 | Bearing the Weight

PROGRESS IN MOTION

When I was in college, I had a couple of stalkers. This wasn't rare. Most of the women I knew could tell a story or two about some odd dudes. There was the one who would appear out of nowhere, literally jumping out of a tree once while I was walking by. Another one I commonly saw at parties who stared too long. At one point I compared notes with a friend, and we realized we shared the same stalker.

I was too nice to these men, having sympathy for the socially awkward, given my own oddities. But even faint attention only seemed to encourage them to come closer and ask for more. As I grew older, I was more careful to nip that in the bud. If you stared at me too long, I looked away. If you followed me, I pointedly walked away. I was trying to avoid long-term awkwardness for both of us, but I was also fatigued by the whole process. I did not ask for this behavior; I did not want the attention.

And so I found myself outside of Boston on my first work trip when I was twenty-two. I was working at a now-defunct start-up, and I was the account manager for a flower company using our website technology. I arrived earlier than a colleague, and one of the client leads picked me up at the airport. This wasn't a romance movie; this wasn't love at first sight. This was work.

The client suggested Italian at a local restaurant. I assumed this was standard protocol, so we went and engaged in awkward conversation while enormous servings of food were plopped down in front of us. He was being very nice, but the conversation sometimes felt dangerous, like it was veering into territory I didn't want to explore, the "did I have a boyfriend" territory. Post-meal, I was tired from travel and happy to get to my hotel. I sat down at my laptop to catch up on work, and then the phone rang. He was downstairs. He asked if I wanted him to come up. I stuttered. I was as polite as I could be, but as firm as I could be: no.

I pinged my boss and relayed what had happened. I was just hit on! He was shocked. I recapped my lack of interest, with the need to somehow defend it. I think I was, even then, intuitively covering my ass. Who knew what story the client would tell? I wasn't angry; I was amused and annoyed. I didn't expect this, but I also was not surprised.

The next day the client was mean and brittle. Previously complimentary, he belittled my work in front of others. He sneered and criticized me. There was clearly a price for saying no.

When I got back to California, I was taken off the client's account. I welcomed this change. Why keep working with someone who hates you? The CFO, the resident adult, called me into his office and made sure I was okay. I thought everyone was fussing. Didn't this happen all the time? I didn't once consider if the incident cost me anything since I was so relieved to be out of the situation. In fact, I thought it was very nice the company was recognizing the situation and addressing it.

This was 2000, and I was twenty-two years old. I didn't think of my career and being taken off a highly visible project. I didn't think of myself as someone representing women in tech. I didn't think I was standing up for anyone else. I was just glad I was never going to have to work with that man again.

Back in the early 2000s, I would attend company ski trips. This was one of the perks I didn't know to anticipate when I joined start-ups. I didn't grow up skiing; it's an expensive hobby, and we didn't have money for the rentals and clothing. Plus my parents seemed to prefer warm weather on the whole; I don't remember many wintertime escapades. I was nervous about these trips since I didn't know how to ski. I learned later, but it was another way I didn't seem to quite fit in with others.

One evening, there was a big party with buffets and a DJ. I walked into the ballroom of the large ski resort, and there were women dancing inside cages raised up on platforms. I sighed deeply. I wondered whether people from HR had reviewed the plans. *At the very least, put some men in the cages,* I thought to myself.

It took years before I stopped seeing some form of slinky women flitting about work events, whether circus types or modern, silver-clad dancers. They did add more men, sometimes on stilts, over time, though. Does that count as equal opportunity?

Fast-forward fifteen years. I'm sitting in a large room at a conference for my team's leadership. There's a fun event planned, a murder mystery. As it begins, I feel my spidey sense trigger—this is about to go terribly wrong. It is a 1920s murder mystery complete with all the stereotypes of the era, including a horny rich man, a slutty maid, and an innocent starlet. Within minutes, the first penis joke was made. I winced, even though it was highly predictable. A woman sitting next to me pointedly got up and walked to the back. Within minutes, the actors visibly toned down their language. At the end, a leader got up and apologized for not vetting the act better. This is progress in motion. Social change takes time, observable tick-tock time.

Last year I noticed a woman leader I worked with wearing her hair in beautiful curls. I was used to seeing her with straight hair. I complimented her, and she shared how she had started styling her hair like that on purpose, having recently found a good hairdresser for curly hair. She said, "Have you noticed how few women leaders have curly hair?"

It's true. Along with the other unconscious and conscious biases we host, straight hair is considered superior to curly hair, especially in US culture. Sure, straight-haired women are jealous of curls, and tons of products promise to pump body into their hair. But if you look at signposts of culture like movie stars, models, leaders, or YouTube stars, straight hair wins. Beach tousled is about as crazy as anyone wants in ads or TV shows. This isn't limited to women; men with curly hair tend to cut their hair short versus sport their ringlets. And certain types of curly hair, especially that of black people, can elicit even stronger reactions, with black people often being asked to change their hair to fit school, work, or other societal rules.[27]

---

27 Mason, Kelli Newman. 2020. "On Understanding and Embracing Natural Hair in the Workplace." *The Riveter*, Accessed April 2020. https://theriveter.co/voice/understanding-and-embracing-natural-hair-in-the-workplace/.

As a woman with curly hair, I've been conditioned by this environment. Also, my frizz-prone hair can be really annoying, especially on days I'd like to appear put together. Straight hair feels more beautiful to me, even though I know others would "kill" for my curls, as I've been told many times. A key example: I straightened my hair for my wedding, the day I wanted to feel my most beautiful.

While I occasionally dabbled with chemical straightening before having kids, now I tend to wear my hair in a braid or bun. When I let my curls loose, I often get compliments, but it's a convenience thing for me. My hair knots more if I wear it loose. I can feel it getting crazier and crazier on humid days, and I'll instinctively pull it back, not even realizing I'm doing it sometimes. It's no coincidence I settled down in the relatively dry climate of California.

Back to my coworker: When she asked me about women leaders, I realized I was still blurring who I was. Typically when I presented in front of large groups, recorded video, or got headshots, I would straighten my hair for the day or the trip. Along with feeling more beautiful, I also felt more confident. I already acknowledged that I looked the part of a poised leader more fully, that I fit in better, when I had straight hair. I was playing into that bias.

But what if I didn't? What if I started wearing my hair curly and showed everyone it didn't matter? What if someone in the audience looked at me and thought, "That could be me"? A couple months later, I was presenting to a packed room of about seven hundred people, with another eight hundred watching over livestream and in overflow rooms. The stage lights hammered down on me. I wasn't nervous about what I was saying, but I was pretty sure my hair was frizzing up in real time. My hands itched to pull my hair into a bun. Afterwards, two women walked up and complimented my presentation *and my hair*. I explained how I was doing it on purpose, and I saw the look of appreciation in their eyes. It mattered.

This series of stories demonstrates my progression as I moved through my career as a woman in tech, from experiencing relative obliviousness, to awareness, and finally to leadership.

In my early years, I had the privilege of occasionally observing issues like female entertainment at parties but mainly being able to move on and believe someone else would correct any problems. I considered women's groups and events to be a distraction; I'd never liked sororities, so why hang with women now? It's not that I didn't have female friends; I just didn't derive much value from the network springing up around me. I had a job, I was successful and moving along, and I was trying to stay on top of both a fast-paced career and personal life trajectory.

But as I became a leader, the ramifications were impossible to ignore. I was responsible for people's salaries, their career progressions, and often their next opportunities. As a senior woman in tech, I also felt responsible for speaking up and acknowledging what wasn't working or what could be better. I recognize now that I am here to speak up where others can't. So in recent years, I've spoken up more than ever before in my career, particularly when I see issues that would hurt our inclusion efforts. The diversity and inclusion conversations, as well as the #metoo movement, have improved my ability to speak up for myself as well as others. Compensation is a key example of where this has worked. I can now speak with my boss and say I'm not happy with my compensation, and I'm doing this not just for me but also for other senior women. As someone who finds it difficult to negotiate for myself, feeling responsibility toward others is particularly empowering. This combined with often progressive policies and benefits in tech, e.g., more generous parental leaves than required by US law, assist my ability to do the right thing.

But this weight can be difficult to bear, and anger can result as well. Why should we have to do more than others? Often the work to support women and minorities in the workplace, like running employee resources groups or going to diversity recruiting events, is considered thankless and goes unrecognized by promotion or compensation processes. There are days when I think we are making progress. There are days when I am staring at the eyes of well-meaning people and thinking it's not enough. Days when I am ranting about how a woman would never get away with what I just saw a man do. These things can all be true. Why can't I just do my job and be successful and not think about my vagina at all? I don't have a pat answer to this challenge, but at my core I'm still the girl who spent hours alone in the computer lab figuring out how to build a website. I found my people there, and I'm not going to give that up. As I interviewed women, I was curious about how they dealt with the weight of representation. Do we ignore, accept, or embrace our potential role? And how do we feel about the path we choose?

HOW IT LOOKS AND FEELS

Most of us have probably at some point felt different or singled out in a room. We may especially remember this happening during childhood or when we were a teenager, when conformity is typically valued way more than uniqueness. What does being rare look or feel like in the workplace? How do we know we are different? Why? In talking with women, often the possible causes are blurred. Was it because they were a woman? Were they too young? Too old? Was it because they were black? Latina? Not technical enough? Didn't go to college? The list went on. Sometimes this didn't bother women, but sometimes it made them wonder what was causing the feeling of not belonging.

Carol is a software engineer manager who studied mathematics and computer science. Prior to moving into management, she was a full stack developer for ten years. "I definitely noticed that sometimes at a meeting, I am the only female there, for example." She'll also see that she's the only female on an interview panel, "and I think the company did it consciously, to make sure that the candidates feel the diversity and inclusion of the panel." These things haven't bothered her much, though. "In a way I feel very lucky. It's a culture that everybody needs to be conscious about, but also at the same time, I acknowledge and accept that's the current environment that we are operating in."

Paola sees it around her and also when other companies try to recruit her for program management work. "I'll look through [their website], and there's no women. It's all men and maybe one woman in HR. And then it's worse for women of color. Even in my team, I couldn't even tell you five people that are Hispanic." She describes a weekly infrastructure meeting that pulls together all the leads of the various areas. "And so I'm the program manager, again the only woman of Hispanic origin. And then I think out of that whole meeting, there's probably two other women, and there's probably eight other director-level [people]—the men." While she won't allow herself to get intimidated, it would certainly be possible given she's "not a director, and a woman of color, and Hispanic."

Leanna really doesn't know how much she struggled in her early career "because I was a woman and how much I struggled because I was twenty-two." She was also less technical than others. "And then I think, how much did I struggle because I was in a room full of older men who had degrees in computer science, and I was a younger woman who had a degree in magazine journalism?" As a strong-minded woman, she sometimes feels like she falls between—is she not "chick enough? Am I not a dude enough?" She gives an example: "I'm sitting in Zurich, where all the men are—all the employees are men—and it's just to meet a bunch of dudes, and they are more technically experienced than I am. And I'm trying to problem-solve with them. And they're like, 'No, no, no, you shush.'" She doesn't feel like this is necessarily happening because she's a woman. "I don't know if it's because I'm a weird chick, or if it's because I'm just not a technical expert."

In the world of business development and partnerships, Kristen is often one of the only women in the room. Her upbringing was completely opposite. "I'm one of five kids. I always played sports. I was pretty good at sports, so I've always played on a team. And I have three brothers and a sister, and I was treated the same way, for the most part, by my parents." When she was growing up, she was never told she couldn't do something because she was a girl, and she didn't "really understand feminism until I got to the workplace, and I didn't identify with it. I didn't think we needed it anymore." With a career in sales and tech, she's now found one of the challenges has been being the only woman in the room. "I think I was once the only woman besides the flight attendant on a flight from New York to CES in Vegas. It was me and the flight attendants; [otherwise,] the entire plane was men. Then I spent a week in Vegas with all men, where the only women were in their underpants serving cocktails."

THE ADVANTAGE

Some women I spoke to brought up the positives more than any burden of being a woman in tech, whether because they were surrounded by women on their team or they enjoyed the advantages of having a different approach or point of view.

Carol gave an example of being in a male-dominated meeting that was running over time. She used a feminine tone to nicely ask people to stay, since it was critical to have the people with context finish the conversation. "When I asked people nicely, people stayed because I asked." She also loves being asked to serve on interview panels so she can demonstrate her company's representation to future candidates.

Leanna has seen similar issues arise. She was talking to a friend of hers who is a larger man, while Leanna describes herself as a "tiny little miniature person." They were comparing notes, and he revealed how he'll receive feedback that he's aggressive when he says something abrupt. "But when I say something abrupt," Leanna says, "I get away with it every time." She gives the example of telling her boss's boss to hold on to his idea because the meeting was running over. "Nobody considers me a negative authority figure. They do for even the slightest infraction from [my friend]. And I think that that is fascinating. I feel like I've got some significant female body privilege, because I can be a little more aggressive, but because I'm physically not menacing, I get away with that, which is so sad, but it's something I notice."

Kathleen also feels that it can be easier for her to be direct. "In some ways I feel like I've been able to survive this environment because I might have some tendencies that are more masculine, and though I'm a minority, I also can pass as if I'm not one." While she has that advantage, she does need to watch out for downsides. "If anything, people might give me feedback that I'm too direct. I always like to take that with a grain of salt. If this came from a male, would I have had the same reaction?"

It's also worth remembering that some teams are going to have strong female representation, a topic that doesn't come up much when we think about tech companies. Marketing, customer service, and legal teams can often have higher percentages of women than men. Caragh's experience in human resources, which is predominantly women, has meant she's never felt the weight of representation. "I feel like I really don't bear the burden," but she knows that women in more technical teams do.

BEING AWARE

That said, the women I interviewed did describe the challenges of being a woman in tech regardless of their job role or team. Many would first bring up a reminder: tech is trying, but it isn't perfect.

Camille reminds us that "tech is like the rest of business in America." While there's an assumption that these "younger, newer, hipper tech companies have got diversity figured out," we're still a work in progress. "A lot of the tech companies are just giant companies that have to run like companies, whose employees are just people who have the same issues as a person working in a bank, a person working in a factory, or a person working at Google, a person working at Facebook. So there's no magic paintbrush [they] can sprinkle on all of us, no fairy dust that's going to make everybody down with diversity, inclusion, and equal rights and all that stuff overnight."

Along with that reminder, Camille warns that we will face familiar and new struggles, even with the help that is out there. "There are people here dedicated to attempting to make big change, but there are human beings that work here, and so changing the mind of individual human beings, no matter what [or] where you're coming from, takes a long, long, long [time], like generations of time." She appreciates being at a company that's making an effort, but it still takes a level of patience and acceptance. "It's great to be here where people are at least verbal about it, but it doesn't mean that things are fixed. So if you're not willing to deal with the fact that it's not la la land, be aware. It's not a bad thing. It's life at a giant company anywhere in the United States of America."

Sara P. echoes this thought. "I think that because of how tech works, people also think that hand in hand will come these very advanced ways of thinking, ways of working together. And that's not necessarily the case, right? Business is business. People are people." She remembers talking to a woman at her previous tech company who had presented in front of leadership. "Her boss's feedback after the presentation [was] she needed to work on presenting more masculine, like really needed to command the room, and that it was too feminine." She reminds us that moments like that will exist and vary; "depending on where you go in your career and what company you work for, it can be better or worse."

Similar to Camille, Sara finds that companies that are tech focused are at least talking and thinking about diversity and inclusion more than the industries her friends work in. "People are vocal and actively thinking about the culture and organization and being intentionally inclusive and creating diversity and diversity of thought and that really being a value prop."

Melanie gave me the perspective of working in a small company where the resources and focus may be different. "Being in tech, most likely the opportunities could be with smaller companies where there is not as much work that is being done to train managers or to really promote people who are being unbiased." She thinks there is a challenge to be aware of there, because "a lot of the work has to be done individually and every time over and over at the company level." That said, she notes that there is a general attention now to these issues that is helpful. "I feel like we see in the era that we're in . . . there is more and more focus for even small companies to create agreeable and successful work environments."

THE ASSUMPTIONS WE MAKE

One of the ongoing burdens that we'll face are other people's assumptions. I've heard plenty of stories over the years about men assuming women in the room are the administrative assistants, notetakers, coffee distributors, or party planners. The effort that it takes to continually reassert ourselves as belonging in a variety of roles can be grating and tiring. At the same time, we don't constantly want to war with people to prove women are competent and high performing, and it's unknowable how many women have been held back by those spoken or silent assumptions.

Alex gives a poignant example of how assumptions could have stopped her engineering career before it even began. She prefaces it with, "It often feels like every obstacle as a woman isn't viewed like an obstacle or a small bump, [but] rather a giant wall with barbed wires on top with a 'DO NOT ENTER' sign." In her senior year of high school, she asked her math teacher whether he thought she should major in political science or computer science. "He had seen me compete in debate tournaments, so he suggested political science as 'a more natural choice.'" While he was kind, and he could have simply thought she was a great debater, the comment stuck with her. "[In] the face of every engineering challenge, the voice in my head says, 'see, this isn't your natural ability.' As if women weren't meant to enter the STEM field. As if any woman in an engineering position is swimming against the current and is simply there to prove she's a feminist."

Camille came to tech through the education field, and she can't necessarily separate what she experienced as a woman from what she experienced as an educator, because "the entire early childhood field is very female focused." After entering Google to build the childcare program, she found it difficult to move to other roles. "I was an operations manager, but you're still a teacher, and so you're just this woman who teaches our kids who stays over there." Now she helps other teachers tweak their resumes to sell their transferable skills. "Don't say *teacher* without bullet pointing all the organizational skills, all the people [skills], all the management [experience], all the parent relationships. All that stuff is actually harder than what most people do."

Camille has seen assumptions play out in other ways as well, citing another example where a group of engineers scheduled time on her calendar to question whether she was qualified to collect their money for onsite daycare. She says, "It was not an illegitimate question," but questions why the assumption would be that she wasn't qualified when she was "hired by the same hiring committee, and it was much smaller then, so the same group of people that hired you hired me." The meeting went well, in part because her father is a private bank president, so she knows quite a bit about money collection, but she questions the approach. "They didn't bring any of the rest of the childcare management or any of my managers into the meeting. They just pulled me in and didn't say what for, and then berated me with questions about financial security and banking." While there was some satisfaction in being able to deal with their questions, it still left a mark. "It was like, 'Ha, you can't get me. But geez, is this how you approach things here?'"

As a black woman in facilities management now, Camille continues to see such assumptions play out. While she feels good inside the company, her work with vendor companies illustrates the ongoing challenge. "When you go out into the field, there's assumptions that female facilities [workers] are the customer service type versus the 'know the bones of the building' type." She counteracts this by demonstrating her skills from the beginning. "So you do feel this need to throw out some terminology right when you walk into a room, [language] I don't actually have yet . . . I go to trainings and I read because I'm anxious to be able to walk into the room and not have that [assumption] or to be able to sort of shut down that assumption quickly."

In particular, she wants to avoid explaining why she's in the room. While interviewing women, I heard multiple stories of minorities, especially black women, regularly having to answer questions about why they belong in a meeting or have a certain role. "I live with that a lot, so I am not always sure if it's race related, gender related, or if it's just in my head." At this point in Camille's career, she's stopped paying attention because she can't waste her time thinking about this every day. "When it's time for me to speak up, I say who I am, and then if somebody is like, 'Ooh,' then that's on them, that they are uncomfortable about it and messed up in their own head, or they made the wrong assumption." This reminds me of a Toni Morrison quote:

> The function, the very serious function of racism is distraction. It keeps you from doing your work. It keeps you explaining, over and over again, your reason for being. Somebody says you have no language and you spend twenty years proving that you do. Somebody says your head isn't shaped properly so you have scientists working on the fact that it is. Somebody says you have no art, so you dredge that up. Somebody says you have no kingdoms, so you dredge that up. None of this is necessary. There will always be one more thing.

Ginny similarly describes a major career challenge as "being a black woman and people not expecting that you're going to be as good." This has shown up throughout her schooling and career. "I can remember going to business school and people were like, 'Oh so, are you on financial aid?' Actually, no. My parents saved. I didn't owe them an answer, but I was like, 'Uh, no.' Sometimes they weren't asking; it wasn't racially based per se, but there were stereotypes. Let's face it." Despite this, she has chosen some of the most male-dominated industries to pursue, like commercial real estate and executive recruiting, where she notes all the big search firms were founded by men. As a six-foot-tall black woman who was a single mother, "Who knew what aspect of me would cause people to either lean in or lean out?" She shares how she wanted to be average in high school and "felt burdened by being different." That said, she was aided by her height and strong voice, often projecting the male-associated strength we expect in an executive's presence. "I realized that probably made it a little bit easier and counteracted [assumptions] on the one hand. I'm sure there were times when people were like, 'Who does she think she is?'"

Kathleen also muses over these distinctions. "A topic that has come up recently in my [HR] organization is the idea of professionalism and the definition of professionalism." Clothing choices came up. Despite tech being a relaxed place where athletic wear is frequently seen at work, there's still the lurking question of attire. "For example, when I see bra straps at work, I don't think that's professional." Lately she's started to rethink some of those norms as she wonders at the source of our definitions of what's acceptable or what's not. Revisiting these concepts means thinking about gender and race norms. "I think about who's thriving and who's successful in tech. Maybe it's easier for extroverts to have their ideas heard and their work rewarded or appreciated." Now she has an eye out for which behaviors we're rewarding and whether those behaviors are actually needed for the role, "or is it because we think that's the right thing?"

Women often employed plain hard work to diminish those assumptions. Amy saw being appreciated for who she is as an uphill battle in her job. "I'm the queer Asian woman supporting the south Atlantic who had the bathroom bill controversy when I first started. They play Fox News in their lobbies, and these are my clients." She always wants to lead with her work, first and foremost. "[Being] LGBT isn't visible, and I cover my tattoos when I'm out and about, so I don't really talk about it. And there really isn't a place to talk about it when you're meeting with clients. . . . and I am truly committed to my customers." By paying attention to their needs, she proves herself to them. "What's going on in your office? What do you need? Do you have the technology you need? Are your laptops working or is your network working? I care about that. So they see that I care. They also see that I deliver it." It's through those long-term relationships that she sees opportunity. "So I win their confidence, and if we develop a relationship, perhaps I get an opportunity to change hearts and minds, because part of it is visibility."

WALKING WITH A SIGN

If Alex could wave a magic wand, she'd change the burden she feels to represent all women. "I don't really want to be walking around with a sign that says 'Women in Tech' or be a unicorn anymore. I don't enjoy the responsibility that any mistake I make represents an entire group of women who aren't meant to code. I don't want every company rejection or hurdle to make me think this isn't what women are supposed to do. And I certainly don't want to consistently have to check if I can count the number of women in the company with just one hand."

Alex has felt this weight of representation since college. "I would walk out of a class in college when I'd see more than five girls, because I knew for sure I was in the wrong class." She's had people tell her, "'Oh, you just got that job because you're a girl,' as if there is a quota that needs to be met and I am just here for that." She also misses having more female friends and coworkers. "I was a teacher for a short period, and one thing I highly appreciated was the number of female colleagues I had."

Nevertheless, tech has some winning points that keep her going. "At the same time, however, there are some great things about tech. I don't think the answer is to complain and whine about all men and leave the industry. There are some truly great people who are determined to help women and diversity in tech. I've had excellent mentors, mostly men, who pushed for my growth and encouraged me when I was at my lowest point. The main reason I stayed in tech was my support system: tech community and friends. When you look past the low moments, it is an incredible place to be. To build something completely new from scratch, to invent, to stand on the shoulders of giants is simply unbeatable. The industry is not perfect, but it is consistently improving."

In the marketing and communications world, Karen didn't feel the burden of gender as much "because there were so many smart and accomplished young women around me." Over time, she did think her age might be a factor. "I didn't make a big deal about my age. People accepted me, and I don't think they thought of age either. And that was great." But she was also aware of not being more accomplished when entering tech, which resulted in a lower-level job when she started. She worked her way up, but "over time, age became more . . . I was just more conscious of it. Not in a way like, 'Oh my God, they're all younger than me.' That passed quickly for me, as opposed to a lot of people I talk to who are still sort of freaked out about it." She is used to it, but "over time, I began to see age was a factor and definitely an unconscious bias, I should say, but one that's there, and I don't think probably still has been really fully addressed."

WHEN DO YOU CHALLENGE THE STATUS QUO?

Even with improvements, Bethanie points out what many of us feel. "I think there's always going to be a lot of gender bias, and the tax tends to be on those that are underrepresented to highlight and share and advocate for themselves, and that is just something we have to deal with." She plays a role in explaining this burden to others. "I shared once with a manager of mine that you come into work and think about diversity and inclusion as, 'What are the measures we're going to take to improve it?' I put my feet on the floor outside of my bed and I know I'm different."

In dealing with this every day, many women feel the ongoing price of underrepresentation, and it's not going to be fixed with a simple hiring or training initiative. "This is something that is present to you in many ways, and also many times may not be an issue, but you're wondering if it is. That tax and that distraction of, 'Oh, is this a women thing? Do I have to speak up here? Are they going to be open to it? Am I going to be seen as someone that's always playing the woman card? Am I speaking on behalf of myself or on behalf of all the other women?' You have to carry all that with you."

Bethanie works to improve the overall environment by giving feedback or sharing data, but she's wary of the pitfalls. "I tend to give feedback in a pretty candid way. I try to get data to share. I try to assume good intent—like, 'I don't know if you realize this, but this may have come across as this way'—and in general people have been really open to that approach. But you do have to be cautious to not use it all the time or call out every time you think there could be something. It's a challenge." She feels like she can speak up because she has a strong work history and long tenure, but she still picks her battles. "A question I ask myself each time is: Is this the battle that's worth it?"

REPRESENTING OTHERS

One way women push to change their environments is through helping others. This is why having women support each other individually or in groups and why seeing more women in positions of power is so important—each woman can in turn help countless others with their journey. According to a study of United States Military Academy at West Point classes in the 1980s (the timing chosen specifically when isolation would have been dominant prior to internet and school policy changes), having more women just as peers can significantly increase the graduation rate of fellow women. In particular, a woman in a woman-heavy first-year group had an 83 percent chance of rising to the next year versus a 55 percent chance in groups where there was only one other woman.[28]

Now that Georgia is on a team with more women, she sees how that can make a difference in representation, especially since the CEO is a woman and is committed to hiring and supporting other females. "I think when you have female leaders at a very senior level, more people come. It encourages additional people to feel like that is a good place for them to work." Georgia thinks this is critical to the business she's in, where

---

28 Huntington-Klein, Nick and Elaina Rose. 2018. "A Study of West Point Shows How Women Help Each Other Advance." *Harvard Business Review*, November 26, 2018. https://hbr.org/2018/11/a-study-of-west-point-shows-how-women-help-each-other-advance.

they have to make policy decisions for what content to accept. "You're constantly making a lot of very borderline policy calls, and who is in the room and voicing their opinion and their perspective really does matter. In a lot of those cases, I do find myself being the only woman or the only person from the south or the only person from a religious background. And one of the big resounding themes from our team is that other people in their own situation feel that way. They're the only Muslim, or they're the only person from a country in Southeast Asia." Having a diverse set of people in the room who reflect the broader world is critical. It impacts decision-making, and it's a responsibility tech bears when we are defining the products of the next generation.

As a leader, Kristen likewise takes her role to hire a diverse team and in turn build a supportive environment for others seriously. "I think diverse teams have the best outcomes, and the research that I've read supports that." She used the phrase "I try to bring it along with me" to describe how, through hiring and referrals, she reflects diversity in how she builds a team. This can mean that her team is an island, though. She notes it can be hard at the peer management level sometimes, depending on an organization, "but for teams that I have the opportunity to build, they're normally fairly diverse."

Amy also sees her role as one of supporting and helping others, and that's a role she welcomes. "Being out at work, being an Asian female in the tech world, people see you performing and doing well. I think I'm doing very well in the organization leading global initiatives. That gives people hope. I have the rainbow flag clearly visible, and we've had interns who walk around and say, 'I'm so happy to see that. It makes me feel like I can work here.'"

Laurie K. talks about how her marketing role has given her a chance to represent others better and think about these issues. "We've had a couple of events and then we realized after the fact from feedback from the audience that the panel at the end was all males." In digging deeper, though, "those were all the people that worked on the product. It wasn't purposeful by any means." This triggered further exploration into why this was happening in organizations and what cultural issues might be leading there, especially since the org's leaders wanted to have diversity. She is looking toward how we make our organizations more personal and emotionally savvy as a way to welcome and retain more women. This is inspired by her personal experience. She initially joined an engineering team through an acquisition at her current company, but finds sales organizations to be a much better fit for her culturewise, due to its outgoing and friendly feel. She chooses that despite some of the more interesting product opportunities being on the engineering side. "It pigeonholes me into this role or this org . . . I've had that conversation with my manager, and she's like, 'Maybe it's an opportunity for you to be the change you want to see.'"

Another woman I interviewed gave an example of noticing a team leadership event with predominantly male invitees because of how it was organized. While each invited leader was allowed to include one more junior leader, only the single woman leader had invited another female. The interviewee decided to raise this issue because it was critical to representation at the leadership levels. Her boss at the time was surprised and thanked her for bringing the issue up. She then received a last-minute invitation to the event, which felt double edged, especially due to personal commitments she'd have to miss. "I looked at my husband; I'm like, 'I have to go. Right?' He's like, 'You have to go; you don't light a bomb and not throw it.'" She booked a flight and changed her plan to attend, but her personal enjoyment was dampened by the burden of representation she felt as the rare woman included. For others though, she was glad to have influenced the direction of the leadership team for that and hopefully future events.

VOTING WITH OUR FEET

There are various places in this book where I will recommend walking away from a job or company. I've personally had to make this decision a couple times in my career due to limited growth opportunity, inhospitable culture, or leadership issues. While staying and representing is a huge part of what we each can do, we also play a key role when we leave. While leaving a team or a company can seem like a cop-out or a retreat, it does send a message. In the 2014 Glassdoor Hiring Survey, 67 percent of respondents, both active and passive job seekers, said that a diverse workforce is an important factor when evaluating companies and job offers.[29] When a team has few women or minorities, it helps others navigate their job search and assess a team's culture. Whether you escalate the issue to human resources or not, you are doing something for yourself and for others. In particular, I want to emphasize leaving when your health, emotional and physical, is being impacted and you can't resolve it via other means. There are better places for you.

Diane shared that a key challenge in her career has been finding a place where she can do well, "because there were a lot of spots where you have to be a certain type to really exceed or succeed." When asked whether that felt gender related to her, she says "for sure." She speaks about the first team she joined after college. Her entire reporting chain was men, and her entire team was men too. "I always felt like they spoke a language that I didn't quite understand, and I'm like, 'What do you mean when you say these words?' And then when I would repeat similar words to them, they would look at me like I was not saying English words." Some of the language was consulting oriented, like *synergies*, but she remembers a meeting where a man referenced "testicular fortitude" to push people toward making a decision. "And everyone's like, 'Oh, what?' And then they started to listen to him when he said that."

Diane also had a classic experience: not getting credit for an idea and seeing a man get it later. She was managing a vendor team in India, and "I had an idea of alternating the timing that everyone would start so that we had more coverage on each end of the day. And I took the idea to my manager, and he's like, 'That doesn't make sense. We're fine with what we have.' And then a month later, some other guy on my team had the exact same idea. And [the manager is] like, 'That's brilliant, we should do that.'" When Diane dug into why things like this would happen, her manager would say that the other person refined the idea a bit more. "I'm like, refine? What does that mean? Can you refine that statement for me?"

She was confused and tried to adjust her own communication, but it never worked. "I kept trying to adjust my behavior to fit what I thought

---

29 Glassdoor. 2014. "What Job Seekers Really Think About Your Diversity and Inclusion Stats." Published November 17, 2014. https://www.glassdoor.com/employers/blog/diversity/.

that they wanted, which I think made things worse." She does think she'd cope better now. "Now I feel like I have the skills where I probably could have navigated it better, but [not] being fresh out of college and having that be my first real job."

She chose to leave, and she found acceptance in her new team. "I found out that people really did care what I think; they wanted my opinions heard. They didn't call me emotional; they didn't think that I was just mimicking the behavior that I saw." She feels like she wasn't really able to flourish until she found good managers and teams. Today, she will even go so far as to research the cross-functional teams she works with. If a team is all men versus more diverse, she may adapt her approach. Those first years taught her what she needs in a work culture and to see "the feedback I received there wasn't necessarily about me specifically but more about how women work in the workplace."

Georgia had a similar wake-up call about how different two teams could be. When she was working on a team that interfaced with traditional telecommunications and construction companies, "I remember multiple meetings with vendors or third parties where I would walk in the room and they would wait for a man to walk in the room from [my company] before they got started. Or if I started to run it, they couldn't comprehend why a woman would be [doing that]." While that didn't make her move on at the time, she now sees the difference in stark relief in her current team. In her current role, she is "shocked by how many female leaders there are and how I really kind of stopped thinking about it."

CROSSING THE LINE

Harassment and inappropriate behavior are critical topics with their own experts. For this book's purposes, I'm simply going to assert that *there is a line*, and when it's crossed, women shouldn't be expected to deal and consider it part of their role. As the #metoo movement has shed light on physical violence and harm toward women, we've all acknowledged the various shades of gray we've been dealing with for years, often silently. Behaviors can include ridicule, name-calling, offensive jokes, threats of physical harm, and more. If you are experiencing workplace violence or harassment, there is support for you. Please see the Appendix for more information.

I didn't collect many of these stories, as that wasn't my intent while interviewing women about their careers. Many women outright said they'd had a positive experience in tech and felt protected by its culture. Others alluded to issues that didn't appear central to their stories, and I didn't ask questions that probed directly. In other words, I didn't ask, "How many times have you been harassed in your career?" That was on purpose; I wasn't writing a book about harassment but rather how women built their careers. Others shared stories with me but asked that I not share them due to the people involved. For all the openness we've been bringing to this subject, it is still an emotional topic, and women still fear and experience repercussions for sharing their stories. As a result, I'll share some anecdotes below but all anonymously.

One woman shared, "I've had to deal with harassment on multiple occasions, and we can't network in the same way. You think about men [who are saying] 'I don't know how to talk to women anymore.' My whole life has been everyone's going out to drinks, everyone's having a good time. This is what we do after work. And then why do you have to go and say something awful? . . . I mean, luckily HR has always been really good. But it is, again, frustrating because the incident happened. You have to decide whether or not you're going to do something about it. And then all these rounds of proving and talking and going through the whole thing over and over again for them to be like, 'Oh yeah, that was a bad thing.' I just want to do my job. And that's, I think, the challenge—you just want to do your job and not have to worry about harassment."

Another woman spoke about how she feels like she's been lucky, giving the example of how both her direct manager and VP are female. She did have an incident that continues to stick with her, though. "I had an incident when I was getting promoted, and I thought a lot about whether I should have reported it then. I was like, I want to get promoted, but I also don't want to blow things up." She received the promotion, and one of the reasons cited was that she received positive feedback from the person who was perpetrating the situation. "I was like, all right, I guess I made the right call. I did what I needed to do to get promoted."

As time passes, though, she continues to look back. "I felt a lot of guilt second-guessing that I really passed the buck, which was good for me individually, but I never really confronted that person in any meaningful way, aside from like, 'Hey, sorry, you know, really not interested, this isn't really cool. There's a conflict of interest. You're writing for my [promotion], I need you to write nice things about me. There's no way this is going to happen right now." She thinks about whether she did the wrong thing, "whether I didn't hold up my end of the bargain being female and in this environment where it is a problem for my own self-interest." She's still torn. "At this point I think it's so long past, and this person still sits over there." She wonders whether to share what happened with him and how it was not okay.

Another woman looked back at a culture in a male-dominated organization. "I felt like I didn't get the seat at the table where I have had it before, and there are a lot of instances where I wasn't necessarily treated with respect. I think that was because I was a woman and also softly spoken, which has its downsides." Beyond that, it was a culture that allowed brutal behavior. She described it as "cutthroat kind of dog-eat-dog world," "competitive," and "stifling." "There were people in teams that wouldn't go into certain offices for fear of being screamed out, and the level of profanity was pretty bad too." Talking to her manager wasn't helpful, as he had also been "incredibly disrespectful, more so to women," and others also seemed to accept the behavior. "They are who they are, and you're not going to change anything and you can't change anything." It was hard to accept, especially given she had a team she felt the need to protect. But ultimately she left the company due to the unacceptable behavior.

Ending this chapter on that tough note was a hard choice, but it's important to recognize the downsides as much as we explore the upsides. As we explore what it's like to be a woman in tech, I'm trying to seek a balance of highlighting the good and the bad. This is how we move forward, by truly seeing things for what they are and building our tool chest for changing and dealing with different situations and environments. Let's tackle a topic next that I find fascinating: likability and women.

5 | The Likability Problem

I was on fire.

Angry at senior leadership, I rapidly drafted a letter expressing all the ways they were failing the team. Later, another woman leader and I edited and reedited it. Eventually the initial language was tempered with cool, calm logic, as it should be before you write your boss's boss's boss. But in the first draft, as I was summarizing all the ways we'd stepped up as leaders, my frustration was clear. We'd done everything we were supposed to and we still hadn't been allowed access to the top, where we could really make a difference. My summary boiled down to one line: "We've been well behaved."

I don't recall thinking of the well-known quotation "Well-behaved women rarely make history"; I was thinking of how I'd been a good girl. I'd done everything I was supposed to—been both nice and successful, backed up the people in power (who were all men in this case)—but they never welcomed me into the higher-level rooms where investment decisions were made. And in those rooms, leadership had been making the wrong decisions for years, as it turned out. I'm not a superhero; I didn't believe every single thing would have been fixed if I'd been there. But not having been in the room meant I was powerless, toothless, and frightfully complacent. I was angry at myself.

Deleting the "well-behaved" sentence was the right thing to do; our letter was welcomed and acknowledged because it was strongly worded but also practical and pragmatic. However, my fatigue with being a good girl has not dissipated. The standards of getting ahead as a woman are conflicting and dangerous. How can I be direct but not too direct? Be nice because no one likes mean girls, but not too nice because nice girls don't win? How about looking professional because appearance matters, even more so for women, but not too professional because Silicon Valley rejects fancy? Making the perfect presentation slides so you'll approve my request, but not looking like I'm good at presentation slides because that's not a real skill and means I waste time? (Or worse, you'll only want me to do slides forever.) Or what about how I have to influence without authority, but if I'm too good at it then no one realizes I'm doing it?

Over and over I ask myself this question: Can you be so good at being good that you become invisible? And when we're invisible, who has the power?

That there is a correlation between success and likability for women is hardly news. Sheryl Sandberg covered this in her book *Lean In*, often quoted both positively and negatively when discussing the status of women in the workplace. Marianne Cooper, its lead researcher, reasserts this in an *HBR* article entitled, "For Women Leaders, Likability and Success Hardly Go Hand-in-Hand." Marianne highlights decades of studies done by psychologists "which [have] repeatedly found that women face distinct social penalties for doing the very things that lead to success." She gives the example of women "who are applauded for delivering results at work but then reprimanded for being 'too aggressive,' 'out for herself,' 'difficult,' and 'abrasive.'"

The "likability" concept has had long-lasting durability, impacting how we interpret women and minorities in the workplace, as especially highlighted in recent political elections. While a man can be successful, be likable, and appear focused on his own personal gain, it is nearly impossible for a woman to tread there given how we define women's roles in both society and likability. As Claire Bond Potter, a professor of history at the New School and the executive editor of *Public Seminar*, asks, "What would it mean if we could reinvent what it is that makes a candidate 'likable'? What if women no longer tried to fit a standard that was never meant for them and instead, we focused on redefining what likability might look like: not someone you want to get a beer with, but, say, someone you can trust to do the work?"[30]

While this trap may seem untenable, it is gameable. We can figure out

30 Potter, Claire Bond. 2019. "Men Invented 'Likability.' Guess Who Benefits." *The New York Times*, May 4, 2019. https://www.nytimes.com/2019/05/04/opinion/sunday/likeable-elizabeth-warren-2020.html.

how to navigate typical societal norms through trial or error, or often perseverance. For example, since women are expected to be focused on the community and fulfill their roles by being so, many women are able to achieve success and likability by making sure their efforts are correlated with the community's needs—in other words, not only personally beneficial but also servicing others.[31] That's why I can feel as if I'm on more solid ground when I argue that my salary concerns reflect what other women's concerns would be, and I must do this for the benefit of the broader community. Joan C. Williams, a professor of law and coauthor of *What Works for Women at Work,* describes how powerful women "flip feminine stereotypes to their advantage,"[32] for instance blending "authoritativeness and warmth," like sharing a personal story before a big meeting so people see you as a person first, or carefully and rarely choosing when to be firm. As Ms. Williams notes, though, the adaptations require work and are not particularly fair to the women who have to do them.

After reading the research and articles, I was curious. What does this look like for women in tech? What are the versions they've seen of the likability trap, and how are they navigating it? Again, while I didn't ask a question specifically about how women experienced this, stories came up naturally as I was interviewing them.

THE LIKABLE WOMAN

If Sherry could change one thing about working in tech, she'd change the need to always be a likable female. "I would love it if there was more room to behave in a range of ways. Like, some days I'm really likable, some days I'm not. But I feel like there is a sense that you always have to fit in this range." She sees this impact her VP, who is also female. "She just made VP [a year] ago, and it's something that I see out of her as well, where it's a very concerted effort. Maybe she's also very personable, but it also strikes

---

31 Heilman ME, Okimoto TG. "Why are women penalized for success at male tasks?: the implied communality deficit." *J Appl Psychol.* 2007;92(1):81 92. doi:10.1037/0021-9010.92.1.81 https://www.ncbi.nlm.nih.gov/pubmed/17227153.

32 Williams, Joan C. 2019. "How Women Can Escape the Likability Trap." *The New York Times,* August 1, 2019. https://www.nytimes.com/2019/08/16/opinion/sunday/gender-bias-work.html.

me that I'm pretty sure she wouldn't have gotten to where she is today if she didn't seem so pal-ly and chummy with all these very senior people." Sherry asks, "What if there [were] a chance for a different personality type to be able to succeed there?"

Her previous role was managing software releases, ensuring that engineers fixed critical bugs before launch, a role that felt "antagonistic to the rest of the team, and I'm a very team-oriented person." Right away it was clear that she was going to have to be very likable "if I ever wanted a shot at feeling like I was part of the team." While successful, she did feel like she was acting out a persona that wasn't her personality. "I don't think it was not authentic, but it was very obvious to me that there'd be no other way to succeed." She reflects on a male colleague who plays the tough guy role, and she doesn't think she could have gotten away with that as a female. "If I wanted to, I think it'd be pretty hard, and I'd probably be half as effective."

Mai has also been there. "I remember I was in an executive meeting once, and I'm a petite Asian female and I was passionate about something. I was probably talking about diversity and how that joke wasn't really funny and how I can't be the only one raising [that] flag, and my CEO was like, 'Whoa, that's a lot. Tailor it back. We can't absorb all this.'" Mai knew why. "It's because I'm breaking the mold of an Asian, female, petite, docile stereotype. Whereas if a dude was to say this stuff, you'd be all about leadership and 'Oh, that was wonderfully put.'" In the end, she thinks it's about their expectations of her "because you're having a hard time aligning my voice and my thoughts with who you think I am."

NAVIGATING EMOTIONS

Jamie also pays attention to how she is communicating. "Especially as a more junior engineer, I would be so desperate in a meeting for people to understand me that I would end up bright red with water leaking down my face. Not crying in a sad way, but I was so frustrated [because] I know what I'm talking about and you're not hearing me." One way she has approached this is to speak less in meetings and communicate more in other ways, like via written documents. In her current leadership role, though, that's not enough, because her teams are distributed across offices. She'll sometimes realize her team members in other locations don't know what she knows. "More often than not, I feel like a crotchety old man—'Get off my lawn!' 'Why are you doing it that way?' 'That is obviously wrong, so obviously wrong, but you obviously must know it.'" She knows people have no idea why she would feel this way, and so she's actively figuring out how to "tune the knob" and optimally communicate, "such that I don't come in as being strong and loud and overbearing, because I can be all of those things."

Jamie calls working on this her "perpetual tightrope," and she now has a couple of team members whom she trusts to give her real-time feedback, for instance at the end of a meeting. "So that I've got something more than just my spidey sense to let me know, because I find that if I don't have those external checks I'll drift, either overcommunicating and having everyone not want to be in meetings with me or undercommunicating and wondering why everyone is being such an idiot—not in those extreme terms, but yeah." I asked her whether she thought this feeling was gender related, and she does wonder "if a guy in my same role would care if he would be perceived as being loud and overbearing or if he would even get that feedback."

She looks back and thinks about her undergraduate class with fifty-five engineers, of which five were women. ("We were easy math, 10 percent of the class.") They grouped together and "made sure that all five of us were always as awesome as we could be," not to be antiboy, but rather as "an easy way of getting a reasonably sized study group." Like others, though, she admits she went into engineering because she liked hanging out with guys more than she liked hanging out with women—"It felt more like my people." It wasn't until she started managing other engineers that she "became much more attentive to the gender inequities that either my people were seeing about me with respect to the other leaders or that I was seeing among my people." Caring about other people's careers helped her become much more aware of the "pervasiveness of some of the gender bias" impacting her earlier in her career. For example, "There was a guy who was a technical lead for a project, and at some point very junior in my career, I knew that he was wrong, and I wouldn't let him steamroller me during a meeting. And he came to my office afterwards and was very shaken and said, 'Are you mad at me?' And I said, 'No, I just thought it was really important for you to understand this, and if I came across as obnoxious, I apologize.' He was like, 'Okay, but are you mad at me?' And after like the fifth 'Are you mad at me?' I was like, 'I'm starting to get mad at you.' And in hindsight, it was like his wife or his mom or whatever was yelling at him, and he didn't know how to respond to me as an engineer. But even then, I didn't quite grok that. It was somebody else saying, 'Yeah, he's doing that because you're a girl.'"

When women try to address this type of feedback, they often can achieve polarizing results. Michelle hasn't necessarily felt singled out due to gender as a woman engineer, but she does think that "the evaluation process for women is so much more strict." She noticed that in the same performance review, one person would say, "Michelle is too abrasive and selfish," and someone else would say, "Michelle needs to speak up more and have opinions and stop putting other people first." The contradictory feedback can often leave women feeling like they are in an unwinnable situation. "I'm like, I can't do both these things at the same time, and so I felt I didn't know how to take often contradictory feedback and turn it into something actionable."

She does notice this is different than how the men she knows approach their careers. "When I chat with all my male friends who are at approximately the same career stage as me, I'm way, way, way more cognizant of my reputation. [For example] who likes me and who doesn't." She doesn't want to have to care, but she's found enough situations where her likability and reputation matter to getting work done.

THE UNLIKEABLE JOB

Georgia felt the pressure of having an entire role that wasn't particularly likable when she was responsible for deciding whom to lay off from a large organization. "I remember this disconnect and really struggling with: I know these people. They're friends, I know these people. It's coworkers. I hired many of them. I've worked alongside them. How do I then go into a room in the back and figure out who actually isn't going to have a job anymore? I had no training in [how] to do that. . . . And I really loved what we had built. It was more than a job to me at the time." She struggled with how to disconnect from the emotional side while making the right decisions, and she felt torn during the experience.

She doesn't particularly attribute this struggle to gender. "I'm very empathetic, and so I feel like I might've struggled with it more than some people did. I wouldn't necessarily attribute that to gender, but if I feel like it took a larger toll on me, then I saw other people who are able to be pragmatic about it, in my view. Whereas to me, I was making choices about people's lives and families and security and separating those two was really hard." That said, she found it helpful to connect with women colleagues who were also struggling to help her persist through this difficult situation. "There [were] enough people involved that I think having a group of people, who ended up largely being other women, who weren't trying to pretend like this was a normal thing to be doing, or that it was okay, or that they were okay, helped me realize it was normal to be feeling or dealing with this in the way that I was. So that the struggle felt okay and normalized." She also found it helped to be in control of the due diligence to make sure the right thing happened, and to be able to advocate for those people impacted by the downsizing.

STANDING UP FOR YOURSELF

Ginny, now in her sixties, looks back and is glad she's asserted herself, regardless of likability, throughout her career. She shares a story of when she was a year into her first executive search firm. She was on the phone with a client with another consultant as well as two junior colleagues. The fellow consultant, a male colleague, "made a point over the phone, and I made what I thought was a complementary adjacent point, and we hung up the phone and he—he had a temper—hung up and said, 'You contradicted me.' And they're sitting there going, 'Wah?' I was totally baffled. He was like, 'Everybody get out.'" She was in her thirties and "pretty confident by this point. I had already worked in three other big companies, so I walked toward the door and I closed it behind the more junior associates, and I stayed in the office and I turned to him and I said, 'I have to say I'm a little confused.' The consultant was dismissive, and Ginny went on to explain that she was new, didn't know what she was doing, and he was either going to be "part of the solution or part of the problem, but I did not contradict you." He apologized, and she's since realized his mood was a reflection of his own issues and not hers. Ultimately he became a partner, ran the office, and supported her in "building out the whole diversity practice and becoming a partner and all that other stuff, and then became the chairman of the whole company. I can't say that we're best friends, but there was a lot of mutual respect."

Her overall summary is, "There comes a point when no, you are not going to play me. You are not." She thinks as women we have to step up for what we will not take, but she also knows this is hard. "Trust me. I've had my invisible, inaudible moments where I didn't speak up, where I didn't feel like it was appropriate." It wasn't even due to being intimidated, but more about what we assume. "You presume that these men, when they say stuff so authoritatively, that they have rule of command and subject matter expertise." She's so glad she's sixty now and can "finally appreciate all of who I am and how different and how valuable my insight is." She reflects back on a saying from her mother: "*You don't have to like me, but you will respect me.* And that is more of what I've always sought than being liked."

DESTRUCTIVE EMPATHY

Olivia shared a potential dark side of tilting too strongly into empathy and likeability. "I'm always too empathetic, especially with people that I manage. I'll find excuses for why they don't get stuff done. I'll do things for them. I'll do things for my team that my manager would never do for me." She describes this as "destructive empathy," because then the team can take advantage of her. It also has a downside in that "your leadership team thinks you're too caring" and not focused on the business first, and she's received feedback in her performance review to develop in this area. As a middle child with a people-pleaser focus, she senses when her team members have personal or family issues. While "being able to see that makes me more empathetic for why they're underdelivering," ultimately the work still needs to get done. "If you're always covering up for your team, then there comes a point that it gets you as a leader in trouble. And it did."

She's actively working on this, but she's not sure if she's doing it well. "In some cases it's by being more, 'less feelings, go with business,' and then I come across more aggressive, abrasive, just numbers first. And so it's always been a pendulum. I don't think I've ever found that right equilibrium of 'judge and mentor' at the same time."

YOU FIRST

I think the most dangerous thing about the likability issue is where it might silence us when we should advocate for our careers. While we are thinking of others, it's still important to think of ourselves and respect our own needs.

Annie loves being able to build things in tech. "Whether it's products, processes, data sets, teams, or people's careers—you get to be in the unique position [of building] out a company from the early days, and express opinions that shift how things are done." At the same time, she gives a strong message about prioritizing ourselves. "On the other hand, and I think this can especially impact women and other minorities, you have to put yourself and your career first. The start-up or early-stage tech company will *generally* not do that for you." Her advice is strong: "If you want a seat at the table during a meeting, take it. If you have an opinion or information around a decision, voice it. Of course, I'm not saying you shouldn't be a team player or prioritize your work around company values and priorities. But within that structure, confidence and taking the initiative are rewarded in start-ups and early-stage tech companies. Do that for your career and your team, but also do that for taking care of yourself when necessary."

Reese has come to a similar realization as she's navigated her career. "I think I was an optimist up through my thirties, [believing] that the professional world is a meritocracy." She no longer thinks that's the case "in a world as complex as ours when it comes to race and gender." Based on her experience, "you do have to learn that it's a game in which your actual capability is just a part of a much bigger equation." She thinks women could be susceptible to missing this factor. "I do think that women, because we simply can be so capable, presume that's the key and that that's the factor." When she mentors younger women, she puts it this way: "The job is not the job, and the job has never been the job." This explains the gap between what you think your job is and what it might actually entail. For instance, there's writing emails or doing presentations to communicate your successes outwardly in order to garner recognition—"That's the part of the job that's not the job. But that really is the job."

After speaking with me about the challenges of having a nontraditional background and now being seasoned in tech, I asked Jennifer T. how she's been dealing with those issues. Following the same school of thought as Reese, she instantly turned into an unapologetic woman of power. "I'm very clean and clear about my boundaries nowadays. I got here by truly understanding my value and working with the best clients. These companies typically prioritize diversity in every capacity; they don't care if I'm a single parent and actually appreciate not only my resilience but the professional experience I bring to the table. You need grown-ups in the room too." As a result of her firm approach, she feels like "it becomes a normalized conversation where you're not having to apologize away your appearance or even that you're older." Instead she doubles down on her leadership and communications experience with multiple organizations nicely and consistently: "Oh, how lucky are you that I have dealt with a [nonprofit] board before and I know what to do!"

Rosalyn brings it back to a core lesson of building relationships, even if likability is a bigger factor for women than men. "A huge revelation for me over the years is that you find ways to form friendships and allies, sometimes in unlikely circumstances, but that's actually what carries you through and helps ensure your success." She gives examples of having a friendly smile in the coffee room for the people "that you don't know really well but also happens to be a working mom." It's been key for her to have a "casual space to breathe and talk about your kids in a way that makes sense and doesn't make you sound like an insane mom when you're on a team of a bunch of twenty-seven-year-olds who have no responsibilities and [travel] every weekend."

Part of building relationships is to help people know you better, so you aren't deemed a ladder climber if you're ambitious. She quotes Adam Grant's studies about how success and likability are "sometimes incongruous, which is really freaking annoying, but then you do see it exists.... Being unlikable but being the smartest person in the room can shut you out of a lot of interesting opportunities to collaborate or [find] colleagues or [make] friends even." As an ambitious person, she's had to think about how to present herself throughout her career. "In the end, it matters what people think, and people care about what others think about them, so you really do have to make sure that you're cautious of this and that you're balancing that the best you can."

Over time, I've silently accepted the requirement of likability. I'll draw some lines—for instance, I don't expect everyone to like all my decisions—but I do mostly aim for popularity. Instead of having it be a tax, I decided to have fun with it. My primary tool? My sense of humor. Long ago my brother, a wickedly sarcastic and humorous guy, taught me that being funny is way more useful than being serious, and I have used that lesson my whole life. I'll often introduce tough issues in a light-hearted way, like, "I guess I'm representing the ladies today!" when I'm the only woman in a room. Will this work for everyone? No, humor isn't the tool for everyone. This is the one that works best for me as I navigate that tightrope of likability and effectiveness. The stories in this book may give you ideas for other approaches to try.

In a similar vein of different approaches and ideas, let's delve into the dark but interesting world of why women don't always help other women.

6 | The "Special Place in Hell" Question

Madeleine Albright famously said, "There's a special place in hell for women who don't help women." That sentence went viral after she spoke those words at a rally for then–presidential candidate Hillary Clinton. She later wrote an op-ed in the *New York Times* describing the reasoning for her "undiplomatic moment"[33]: "In a society where women often feel pressured to tear one another down, our saving grace lies in our willingness to lift one another up." Concerned about the tone of the election toward Hillary Clinton's candidacy for president in 2016, she wanted to speak up and remind us of what's at stake. "What concerns me is that if we do not pay careful attention to this history, the gains we have fought so hard for could be lost, and we could move backward. I do not have a magic formula for how every woman should live her life, but I do know that we need to give one another a hand." Even suggesting this idea was considered unlikeable and required clarification, reemphasizing the difficult terrain we face as women in power even as we try to help.

HUMAN NATURE

Over lunch one day, I told a younger colleague I was writing this book. She leaned forward eagerly and asked, "So are you going to talk about how women won't help women?" It wasn't the first time that question had come up. In fact, I already had this chapter planned because I'm fascinated by this topic.

Why? At first blush, we often talk about this conflict as if it's salacious and full of intrigue, catfighting on an intellectual level. But I find

---

33 Albright, Madeline. 2016. "Madeleine Albright: My Undiplomatic Moment." *The New York Times*, February 13, 2016. https://www.nytimes.com/2016/02/13/opinion/madeleine-albright-my-undiplomatic-moment.html.

it disquieting and insidious. Movies often depict this as an almost sexy confrontation between two rivals. Think of *Working Girl*, a classic eighties movie with shoulder pads and corporate intrigue. Scarcity abounds! Melanie Griffith's character, Tess, and Sigourney Weaver's Katharine are competing for the same man and apparently the only position of power for a woman in a large finance company. There's sexism, harassment, the theft of ideas and somehow clothing, but in the end supersmart and supersweet Tess is triumphant and knocks the other woman out of power, garnering herself a corner office in the process. The last scene shows her paying it forward and being kind to her secretary—go Tess! And see how the likable woman won? The reality is far less entertaining and quieter: women who don't help women are often silent, simply absent in battles or not lending a hand when we most need each other. Why? As usual, there's more going on than simply women being mean to each other.

At our core, we're battling the very thing that makes us human—our instinct for survival. In places where positions for women are truly scarce (e.g., on a company's board of directors), there is both perceived and real scarcity. Fundamentally we are animals battling for survival when there are limited resources; we fight for our position of power in order to survive. Per Ching, who referenced her own career in finance, "There was only going to ever be room for one. So if there were a few of you in a room, you're all vying for that one spot to represent. It's not like they open all of them up, so yeah, you're going to get trampled."

It would be nice to say that women should always help women, but when there's scarcity, perceived or not, we are animals and we compete. Why in the world would a woman help another woman if she thought it would potentially mean sacrificing her career goals? Putting two men in the same situation, we would expect the men to pull swords on each other. With women, though, we expect niceness and support, so it's shameful when we won't help each other. This goes back to the likability bind from the previous chapter.

Should we be helping each other? Of course. We should do it because the only way we get out of the scarcity game is by shunning the zero-sum game, helping each other, and strategically changing the numbers in the room. When we study when there are enough of a particular minority in a group to reach levels of acceptance, exact numbers vary but often indicate 25 to 30 percent is the tipping point. Prior to that, the minority tends to bear the burden of stereotypes that are associated with that identity or are used as an example of that minority (a token). Once the threshold is met, then they stop being representatives of that group and are judged more on their own contributions.[34] Camie gave me an example. "As a software engineer I worked on two teams where women engineers made up more than 30 percent of the team, and one leadership team where they were more than 30 percent. In those cases, I felt more a part of the team without the typical microaggressions that typified environments where I was the only woman."

This is how we change the opportunities we have, by helping each other change the numbers in the room. This is how we change the system. Sometimes we will need to rise above our inner instinct for self-preservation, and I'll admire women every time we have the energy and patience to do so. This comes at a cost, though, and I will work hard to not judge on the days we are too tired to follow through. We need to be generous to each other in every way, not just when we can help but also when we can't. That's the way we make change happen.

In the "Our Champions" chapter, I focus on both the men and women who have made a difference in our careers as individuals. In this chapter, I focus on the dark side and what the women I interviewed experience, hope for, and need.

BAD BOSSES

Diane and I spoke about her former female boss, who was a "strange mix of aggression and being passive." She gives an example of getting positive feedback all quarter. However, when Diane received her performance review, it contained feedback about how she had to deliver more on projects and her communication style with leadership. These were both specific

---

34 Schaefer, Agnes Gereben, Jennie W. Wenger, Jennifer Kavanagh, Jonathan P. Wong, Gillian S. Oak, Thomas E. Trail, and Todd Nichols. "Insights on Critical Mass." *Implications of Integrating Women into the Marine Corps Infantry*, 31-42. RAND Corporation, 2015. www.jstor.org/stable/10.7249/j.ctt19gfk6m.12.

types of feedback Diane would have expected to hear while they were happening, not in retrospect.

She felt blindsided by "very basic career-developing items, if they were really as big as she said that they were." A similar case happened where Diane said something in a meeting that seemed misinterpreted, and again she only found out after it ended up in her performance review. "It's like, if you're going to be my manager, really help me navigate my professional career." Instead, this boss appeared to simply want these items on Diane's permanent record.

Diane formed a theory that this boss "was in the consulting world before, and she had to take on a lot of male traits that may have been contrary to who she actually was." Kris had a similar thought when we spoke. "I've met a lot of women in my life, and recently some women that are more senior than I am that are in tech, and I find that they try to close off that [empathetic] part of themselves as a way, as a mechanism to be successful in a man's world. And to me, it seems like you're closing out something that's your total superpower. Don't do that. We don't have to be men to get ahead."

While Diane has sought to understand the roots of her boss's behavior, the takeaway was that her boss "was one of those females who didn't want other females to succeed." And that was "the most tricky, because you trust them and you're like, 'Well, you know what it's like to get here,'" but then the manager did not have her back. Diane now emphasizes really understanding who you are putting your trust in and not expecting something simply because the person is a female.

THE UNINVOLVED

Amy Huang hasn't experienced active backstabbing from other women, but she has a friend in government administration who was told to pick up her woman boss's laundry and children or be fired. While there are bad actors out there, Amy has mostly experienced a more benign form of unhelpfulness as she's been forming women's groups: women who aren't active and won't participate. She's blunt: "I don't have time in my life to even deal with or think about women who aren't going to help. If they're not going to help, good-bye. Thank you. Good luck."

Beam is a design producer at a large tech company. When I asked her if she's had mentors, she says there are "more senior women in my career that I felt didn't go out on the limb for me, didn't look out for me. Even though for some reason I felt that I could trust them a little more because they were women that I talk to." She recounts how she was reorged into a role that wasn't a good fit for her skills. She proactively spoke with her director about this and asked if there were other opportunities. "She seemed to have my back the whole time, but it turned out that she wasn't looking out for me at all." Due to a downsizing, her position went away, "and she didn't stand up for me at all to help me find something else." This was particularly disappointing because her director was a longtime senior woman in the company with connections.

Beam was fortunate that another team was interested in her and proactively altered a role to allow her to be hired, but the experience changed her. "I've been wary about who I trust these days, and I can't assume that another seasoned woman in the industry is just going to look out for me. I mean, that's a sample of one, but if you're not going to show that you're going to help other people, what are you doing?" This has made its mark on Beam and her overall trust of others and the culture around her.

THE QUIET TYPE

Olivia similarly noted that a lot of what she experiences is subtle. "There's a lot of . . . I'm trying to find the right word for it . . . there's definitely microaggressions." She remembers performance feedback she received from a peer that said, "'Her communication style is not my cup of tea but somehow it works for her,' and it was supposed to be a compliment because it was in the areas of strength." But it didn't seem like a compliment to her.

This dynamic has also interfered with Olivia's searches for new roles. While applying for jobs, she'll sometimes see that she would be at the same level as her manager, who is also a woman. "You try not to be intimidating because you're both at the same level, and you think you weren't intimidating, but then you still don't get the role. And you wonder if it was because that person saw you as a competitor since you're both senior level, and you don't know if it's because you're threatening and intimidating. Or if it's because you didn't get it because you shouldn't have gotten that role and there are better candidates."

Kerry experienced a tough culture at her previous company, "with guys that would smash their fists on the table," but she actually found it harder with some of the women. "What really surprised me was one high-up VP attorney. She was the one that people were afraid to come into her office because she screamed bloody murder. She had no time for anybody, no time for women. She'd make offhand comments [like] 'I don't want any interns. I don't want to train them. I don't care.' She really just didn't care about anyone but herself." For Kerry this was hard to see, because "you should be supporting these women in these groups and protecting them and mentoring them. Even if you don't want to be a one-on-one mentor, there are ways to do it without having to meet someone every day."

A QUICK PAUSE

Can you see why I find this area so fascinating? Even from these last stories, the way we think about each other, *as women wondering about women*, is filled with unspoken questions. The temptation to explain why the person may be acting this way is high, ascribing either bad or good intent. In a very "one bad apple can spoil the bunch" way, one toxic woman—even if the exception—can leave a lasting impression, tainting other women. That's a hard place to be for all women, and that's especially true in an industry where there are fewer women overall.

The dynamic between women can be fraught with expectations and nuances, and some women simply find men to be more straightforward. An example from Olivia: "I mean, there are still men that are passive aggressive, but it feels like most of the time you know where you stand. You did it or you didn't do it. I disagree with you and here's why, but let's still be friendly in spite of our disagreement." Rawan had a similar take: "It's been a very interesting trend to feel like men are often a lot more supportive than women in my particular case. And I realized that actually working and reporting to men has been a lot easier and a lot more straightforward, a lot more transparent. Easier for me—to be very honest—than not." The examples she gave were both larger relationship issues (e.g., "the person treated me like a servant") and subtler ones, like how to handle disagreements or apologies. For instance, Rawan spoke about the feeling that apologies for mistakes, even instantly recognized mistakes, were not enough in her communications with women, whereas the path to correct any errors with men seemed simpler and easier to navigate in her experience.

THE LARGER DOWNSIDE

The dynamics in our relationships with women could make us shy away from working with women, even though we're women. That tension played out in Rawan's story. "I joined a team that was all about 'we have the tech lead as a woman, this manager is a woman, this other manager on the crossfunctional side is a woman.' I was like, 'That's great!'" Once she joined the team, though, she found people to be at silent odds with each other even though she wasn't trying to compete. "Everyone had two faces, like nice and welcoming, but then send this email like, 'Did you really just write that?'" This made her rethink, and she now doesn't consider women-dominant teams to be a selling point. "I've interviewed tons of teams, and if the manager is a woman, I'm like: Do I really like you? Am I going to really like you? Now looking and reading every word in their emails, and did they respond or not?" She knows it's a reaction she has to women and wonders if she makes excuses for men. Regardless, she's had bad experiences, and they left their mark.

TAKING THE LAST SLICE

As an advisor to start-ups at her VC, I wondered what Hillary's perspective was with her view across the industry. She's heard women say their best managers have been men and that "the women are the problems." That hasn't been her experience, but "I have had far more male managers, and just because of that, they have been great. I think I've had more great male managers because of the numbers. It's not necessarily anything about women being bad managers."

We reflected whether this would still feel true if the percentage of women in management roles in tech was higher. She wishes more women in power positions would "identify future star performers that haven't already made it big" and focus on their growth. Often women on a successful path end up being highlighted repeatedly by their organizations (e.g., with project opportunities and recognition), while other strong performers may stay in the shadows. "I just see over and over again these very successful, very powerful women being highlighted, which is awesome; they should be. But I also hope that those women help others that don't have access to this pie-in-the-sky mentor or all the opportunities—to help guide them into knowing how to speak up for themselves, position themselves, and to really let them shine and show them the way instead of always promoting or highlighting the women that already have made it. I think there's an opportunity there that is not necessarily being met."

I link this concept of creating opportunities for all women that Hillary raised: "I still do think that women need to be reminded that there's not a piece of pie that if you take one slice, there's less for other people. We're all in this together. I don't know that it's necessarily just from women, but I think that women do need to really look out for others, for other women, and without feeling like by helping them, it's taking away from your capacity and your power in any way. Opposite is true. By giving back, it only amplifies you and your position."

For me, this gets to the crux of what's needed in relationships between women in business: we don't need to be nice to each other or even like each other; I think that's a distraction after thinking about this in conjunction with likability. We do need to respect each other, however, and we are missing out on opportunities to advance the industry we're in and our whole gender if we don't help each other. When I am judging other women for anything—how they dress, how they are carrying their baby, what they are feeding their children, how late they work, how early they leave, whether they speak up in a meeting, what they look like on Instagram, and so on—I feel played. I feel like society has tricked me into trapping both me and my lady companions in some sick game where we're doing the work of holding each other back.

So all in all, I don't think we go to hell if we don't help each other. I think it's worse: we make our own hell right here. Let's help each other. As Hillary said in her interview, "There's a lot of potentially powerful women that need some love and need a little bit of extra encouragement." And I bet you could be both the receiver and the giver if we start now.

Amid all this work talk, I was also curious about our home life. Did the ladies feel like they had one? Or is tech as terrible as reputed with long hours and 24/7 demands? Do they feel supported outside of work, or are they stressed from spinning too many plates? It's time to get personal.

7 | Life and Family

Growing up, I was clearly rewarded for good behavior. In fact my value felt almost entirely external. How I presented myself, the grades I received, and later the promotions I garnered and the husband I married—all were visible signs I was successful. And then I started having children, also a success. It's 2020. Women have thriving careers and booming businesses galore. Why then did I feel like the world was abruptly done with me once I bore children?

There's a hideous photograph of me taken after the long labor and then C-section of my first child. I'm semiposed, mostly slumped over, with my husband and baby. My left eye is sliding down my cheek, troll style, from exhaustion. I told my husband to delete it, but he shared it with the newly minted grandparents, and now that photo has been in multiple family photo albums. When I asked my mom why she included that photograph, she hadn't even noticed that I looked awful. I was officially not the point.

At work, I returned from maternity leave to my previous role. Everything settled into a rhythm where I was doing work and being decently rewarded, but I was in a lull where I didn't have a sponsor helping to propel or advocate for me. That took years to rebuild after my previous sponsor left and then again when another sponsor left. Without that support I looked around, and at some point the world seemed to silently agree I was "good enough"; I was where I was supposed to be. It had wiped its hands of me.

But I wasn't done. I was only thirty-three! And I think this is when I only truly started to take care of myself and set boundaries at work, once I had children. Everything crystallized for me—I had the rest of my life to fight for; it was all my choice now, what happened, and no one else was going to fight for it if I didn't.

While I was writing this book, I kept adding to a document titled, "Writing a book with kids." I wanted to remember all the little moments not only because they were funny, but also because they made me proud. I was happier writing the book with them around, having them see me do it and ask questions about the interviews and the women. My eleven-year-old showed a particular interest in who cursed the most during the interviews. I've captured my four-year-old screeching on interview recordings as women asked, alarmed, whether I needed to do anything about that noise. My transcripts have sentences like, "Hey, that is the kind of package that melts when you microwave it," and, "My daughter was going to try to steal my phone." There were times I stopped interviewing to get a hug, and times I asked the kids to "keep it together" while I tried to listen to someone. And my personal favorite thing I said to my children: "Obviously this is going to go in the book—when you walked in with a teepee for no good reason."

One night, as my son decided he really wanted to fall asleep on the couch beside me while I wrote, I decided this was how it was supposed to be. There wasn't a better way to do this without children. I didn't want retreats where I could be in solitude and power through writing pages at a time. I wanted to be present for them, to write this book with them next to me, always remembering that I am a mother and better for it. I can't overstate the impact my children have on my life, not because of all the whining and the poo, but because they made me realize what life was for in the first place.

I have decided that my concept of "having it all" is really over my lifetime. That I don't expect to be able to have it all now and certainly not ever in a perfect way. I will be deliberate and think of my life in phases. I will never try to have highly crafted birthday parties for my children or handmake costumes unless I really want to. I will be giving and nurturing to myself as much as I am to others. An annoying meeting won't ruin my day; it will be over in an hour, and I'll leave it behind. I'll move on because I have to.

When you sit in rooms with women, it doesn't take long for this conversation to emerge. Whether women are married or not, have children or not, have extensive life commitments or not, whatever it is, we are never far from the topics of well-being, of balancing work and life, of sustaining high performance. While these topics are not specific to gender, many women feel they face a higher burden, a higher tax, trying to thrive at work and in life. We still bear most of the children ourselves, play a major role in running the household, and face society's expectations of taking care of others. And we bring home the bacon: 41 percent of mothers are the sole or primary breadwinners for their families, earning at least half their total household income.[35] And despite the joy and pride all this brings, many women are tired and seeking a better way.

When I spoke with women, I wanted to explore the different paths their personal lives took and how much of that was shaped by work. What compromises or sacrifices have they faced? Have they felt supported at work? At home? I put aside the question of "having it all" for a moment. To whatever extent I could, I wanted to understand how they were doing and how they felt. It's worth noting that this could be a whole book given how women feel about this topic, so consider this a quick tour as I seek to understand this component within the larger story of women in tech.

THE SHEER WORKLOAD

Adrienne has three children and is senior in her career, and she spoke about the pull between home and work: "This dichotomy of wanting to continue on this trajectory that's maybe expected. . . . But then I've got three kids that I care about deeply and want to spend time with, and then there's myself and my husband, whom I care about deeply and want to spend time with. And so I find the struggle of how do you set a career goal that also allows for all these other things that have to happen outside

35 Glynn, Sarah Jane. 2019. "Breadwinning Mothers Continue To Be the U.S. Norm." *Center for American Progress*, May 10, 2019. https://www.americanprogress.org/issues/women/reports/2019/05/10/469739/breadwinning-mothers-continue-u-s-norm/.

of your career." She wasn't prepared for how hard having both roles would be. "I feel like I never understood—my mom certainly didn't talk about this a lot with me growing up—what it means to be not just a parent but a mom working and having a high-powered fill-in-the-blank tech job." She says this even having a husband who does a fair share of the work. "Look, I have the best husband; he probably does 60 percent of the work." She wonders: If she didn't have that, how could she handle the mental and physical load? "I can spend two hours in the evening cleaning up, and then you're expected to do all these other things at work and then thrive and lead a team or whatever it is you're doing."

While we talk about the glass ceiling and bias, the sheer workload as a barrier is often underestimated and "not talked about enough," as Adrienne points out. Even if we improve parental leave, for example, which we should do, the expected and ongoing role we play as caretakers and childbearers continues. "I'm dealing with all these things that I think create barriers in that way." Women feel that we're responsible for the household, whether the kids have enough socks, clothes, and their homework done, and men don't have to do as much. This is what makes us fantasize about having our own wives and wonder what it's like for a man who has a stay-at-home wife. Are they able to simply get dressed and come to work? While we tend to know more people with stay-at-home wives, Adrienne only knows "less than on one hand" women who have a stay-at-home husband. "That creates a whole different dynamic."

I shared with her a story of how my husband, who owns his own businesses, was staying at home with our youngest when daycare was closed for three days. I'm the one who usually takes him to daycare because it's on the way to work. That morning I got in the car. I wasn't carrying seventeen things; I just had my bag and my cup of coffee. I sat down in the car and went straight to work without any stops. *I'm going to be at work early for once, not in a rush and with a clear mind,* I thought, and, *For some people, this is what they always get to do.*

FEELING THE PULL

Sara is single without children but hasn't necessarily found it easy to find balance either. She felt pulled in two different directions, with an interest in both a tech career and a writing career. She used to deal with it by alternating jobs, working in tech for a while and then quitting and writing for a period. "But it always felt imbalanced." Now she's trying to alternate her focus while doing both. "Then the challenge becomes how to bring enough of myself to both of those while still having enough time for the rest of my life."

Having recently bought a house, she's feeling the stress of wanting to do a good job and be successful at the same time she's thinking about settling down in the future with a family and children. "I have barely had time to clean, let alone meet somebody, let alone . . . then I spiral into all the, 'Oh my God, I have this amount of time to figure all this out.'" Ultimately she wants her life to be "livable and meaningful," but she's still figuring out how to achieve that and whether she actually needs anything else to be happy. At the same time, she knows how easily she could let herself be sucked into only working. "It's not just one career, it's two that are now competing with the rest of my life, and how do I make sure that those two things together don't take up 200 percent [of my] time."

RETURNING FROM LEAVE

I've taken three maternity leaves, and I actually split the last one into two parts by personal choice. In essence this means that four times I had to say, "Hold on, career, while I birth me some baby." At the most supportive of workplaces, the hopefully four to six months—if you have a generous leave, less time if not—shouldn't impact your career. But . . . no matter how you cut it, we have to prepare our coworkers to survive without us, hand over our responsibilities and knowledge to others to caretake, and go take care of other business. What they do while we're gone is not in our control. And what the world is like when we return is also a big question mark. I see many women discuss and worry about their returns from leave for this very reason.

Caragh returned from her third maternity leave and found a very different world than the one she left. She had been managing a team of fourteen, but when she returned, the decision had been made to disband the team. "I had a clear opinion, which of course at that time I feel people discounted because they thought I was too close to it to have an objective point of view, which I also found a little frustrating." It wasn't lost on her that the conversations picked up and the decision was made to decentralize her team while she was gone. "They let the team know that this was going to happen, and then I came back and then they had me come up with the plan for how we were going to do it." While she accepted the task at hand, "there was an ickiness to it," given the decision happened while she was gone, the underlying assumption was that she couldn't be objective, the announcement came without a plan, and "I just had come back from maternity leave, which is stressful in and of itself."

Looking back, she can see it brought her toward other opportunities. At the time, though, she had to tackle the emotions involved. "It was really tough for me to get over my own ego, because I had identified as a lead. . . . I was a strong people manager. I can say that confidently." While she could objectively understand the business call, a part of her felt like she failed in going from a management role to a nonmanagement one. She also felt like the perception of her shifted while she was out, in part due to the woman who'd temporarily taken over the role and potentially scapegoated her. "I feel like, unfairly, my manager at the time had this new perception of me that wasn't true."

Thankfully, shortly afterward, she was plucked for an opportunity to help a large group of employees find new roles, leveraging her background in what she calls the "most rewarding experience of her career." At the same time, she has taken away a lot from the overall experience. "It was definitely a time in my career where I feel like there was a disproportionate view" of her work self. Despite feeling happy outside of work, "I couldn't sleep, and so, from an anxious perspective, that was hard to navigate." The job to pivot her team and then also the new role really helped her recover. She navigated this as she has in the past, by "leaning into where there's a business need and where there's opportunity and trying to do the best work there."

THE DECISIONS

Ching made a major career pivot, in part fueled by impending motherhood. She found out she was pregnant with her first baby as she was mulling a job change. She shifted from a fast-rising role in pharmaceuticals to an administrative role in tech, and her feelings about that change have become complicated over the years. One of the biggest challenges in her career has been "deciding to way downshift and become an admin. I think that has actually put, in some ways, more obstacles in front of me now that I'm deciding to restart in the last few years and get back to a place where I feel more relevant." By becoming an administrative assistant, she feels like she is now battling prejudgments about her capabilities. Funny given she also shared that prior to being able to interview for the roles, "as an admin, you had to take the logic portion of LSAT."

Motherhood tied closely into her decision-making. "I think where the guilt comes from is because I really thought I'm not going to be able to switch jobs, have a career accelerate, and have the flexibility or the freedom that I needed." Part of this was triggered by her experience in pharmaceuticals. "You were usually the only woman on a product team, and their stories were super tough, and wanting a family on top of that seemed like, why add another layer of difficulty to your life?" So she intentionally made the career pivot to admin in order to learn a new industry and grow her family. "I don't really believe in work/life balance (I'm going to use that term because it's the vernacular everybody uses). So I thought at that time, becoming an admin would allow me some time for a few years to focus more on my family and give me that space while I'm still able to learn . . . before trying to accelerate, if that ever became an option."

Ginny "didn't become a mother until I was thirty-eight, and that was a function of not having found the right guy." Then life took many twists while she was pregnant. "My mother was diagnosed with lung cancer when I was six months pregnant. I joined Spencer Stuart [an executive search and leadership consulting firm] when my son was one year old. My mother survived that year, and then she passed away a month after I started at Spencer Stuart. Six months after that, I asked my husband for a divorce. So I had a one-and-a-half-year-old baby, my mother had just passed away, and I was getting a divorce, and I had just changed careers and gone into one of the most competitive search firms in the world." She dug deep to hold herself together with little support or understanding in her workplace. "I was the only woman with a child who was a consultant [there] for a really long time. I'd get to work at 7:30 or 8:00, and most of the men had been there since 6:30. So I looked like a laggard, but I was like, I'm not going to let that bother me because I was up at 4:00 a.m. working from home until the nanny came."

As a single mother, she bore the responsibility for parenthood and her career, and Ginny speaks of her journey proudly. "It's always been a source of pride as a woman. I think as women, this is what we do. This is our part of our legacy. We raise our kids, and the challenge is to not allow it to embitter you. These are choices that we make." She learned she couldn't put herself last on her list of priorities, doubling down on meditation and faith to help her through this time. But she admits she didn't always do well, remembering times where she had to take care of her kid and survive on four hours of sleep.

Similarly, Tieisha was propelled by taking care of her child while still early in her career, and her decisions have influenced her viewpoint. "So I was nineteen when I had my son, which is one of the reasons why I didn't go to UCLA and I ended up going to Devry. Because I had to pay for it, I had to work, I had to do what I had to do, but I knew I still needed to get a degree." This experience has given her perspective that others may not have. "I understand things a lot differently than [other] people do because I understand not knowing where your meals are coming from . . . I feel like I have that empathetic gene woven into me because of my experiences." Overall she views this as an advantage. "Some of my counterparts may not understand certain things, so I do think that my upbringing, or being a little different or having to take a different route from my education, really does help me, because we're in this age now with diversity and inclusion." Her background has allowed her to ask pointed questions, like where her company recruits from and whether they are open to different forms of education than four-year universities. "I said to my boss, 'You inherited me, but if you were to look at my resume and interview me, you probably wouldn't have hired me because of the school

I went to. And look at how I perform. Stack me up against Harvard. Open your mind. You know you have bright people all over the place.'"

LOOKING AHEAD

Georgia is now where Ginny and Ching once were and is starting to think about what her career should be while having a family. "I am pregnant with my first, and so I'm really thinking a lot about how to make both career and motherhood work in the future." She thinks her career will remain in tech because she likes the flexibility it supports—of going to the doctor when you need to, of arriving late as needed, of taking calls from home. "What exactly that looks like, I don't know. I think as I learn more about parenthood and what I need to do in order to kind of have that healthy balance, I'll figure more of that out."

One thing she's wary of is a job that requires her attention 24/7, even if the job is flexible. "This is always my joke about how much I love the flexibility; there is a dark side to it, which means because it is flexible and you can work from anywhere, you can work from anywhere, and so you're always on in some sense. My joke with my job right now is that I have to tell somebody if I'm going to be on a plane for more than two hours, because truly being MIA for two hours, even on a weekend, is incomprehensible to people."

She looks ahead and thinks about the obligations of both motherhood and her job. How will she be able to protect her time for either? "I certainly hear women talk about it more in regard to kids. Because it's much harder to keep time sacred with their children." While this can also apply to men, "I don't hear men talk about it as emotionally as I hear women. I think not being able to know that you can shut off and go home and cook dinner and really be present, I have heard a lot of women struggle with." She doesn't know yet what this will mean for the future, but she's thinking about it now in this time of her life.

Similarly, Carol is thinking about what happens if she has kids, and wonders if she would quit to spend more time with her child. This is a call she thinks she needs to wait to make, though. "Usually I like thinking about the future and trying to plan around it. [But] I realized that it doesn't matter how much I plan because I'll follow my heart." For example, she knows an engineering director who has always been a very career-driven woman. "It changed after she had a kid. She decided to go for smaller opportunities so that she could spend more time with her kids." Carol wants to wait and see how she feels, noting that once she's in that situation, she'll figure out what to do.

FLEXIBILITY

While some may wonder if tech is flexible, others are already taking advantage of the flexibility it does offer. Caragh compares her job, which has long hours and flexibility, to those of her friends in fashion. "They're eight to six, but they can't work from home." She appreciates the flexibility that tech offers but notes "because it's so fast-paced, it's like there's no off, like there's emails still coming in throughout the night, vacation, and things were piling up." She warns "you have to be good about setting your own boundaries, because if you don't set boundaries for yourself and set them as early as you can, it's hard to peel back, because no one's going to tell you to stop and there's always going to be more stuff coming in."

Camie feels like tech has been a tremendous career path, offering her both financial independence and the flexibility she needed as she was raising children. "My kids are in college now, but when my first child was born, I was still a hands-on developer. I was going to take the four months off for the maternity leave that they gave, but she just slept, and I was really bored." She called her boss and did part-time coding while her baby slept. "I was still there watching her, but I was able to do something that kept me sane." She also chose to live close to where she worked. "I drew a commute circle around my house at fifteen minutes, and I was always able to find a job that allowed me, when my kids were small, to be within a distance that I can pop into their play at school or whatever and [still] get back to work [quickly]."

WORK/LIFE BALANCE

In these discussions, the topic of work/life balance or well-being is never far away. Christine would love to change how we think about this in tech, especially because she knows people come to the industry straight from college. "You're used to staying up late . . . [y]ou're used to doing work whenever you want. You're used to falling asleep in class." That's not a random example for her. "That was the hardest thing for me when I first started working, to not have time to sleep in the middle of the day. I was consulting, and I had a corporate apartment ten to fifteen minutes away from the client site for my lunch hour. I would drive back to our apartment, sleep for half an hour, pick up food, and eat on the way back so that I can have that nap in."

The topic of work/life balance is a relevant topic for everyone because it speaks to our long-term health and ability to sustain, and as Christine notes, "it's less about the hard requirements of work necessarily, more about the industry and what people are used to and possibly how people are rewarded." In a world where we are rewarded for launches and getting stuff done, we may sacrifice our well-being to hit the deadlines. "All of us are horrible at estimating, so the way sometimes you have to get stuff done is to stay up [and] work after hours and whatnot, and sometimes at the expense of friends and family and health." She wishes everyone in tech would prioritize their health, even though she knows it's hard to do and that often "you're eager to prove yourself."

Frances struggles with this concept of work/life balance because she passionately loves both work and her family life. "Before I got married, I could work as much as I wanted. I was the job, and I loved it." When she started to date her now husband, she warned him, "You need to make plans because I'm busy." She was clear with him that she was focused and wanted to "continue to be able to build my career." She went to business school for a "really intense executive program. I had a great time, and then I got done with school and I was so fired up, I went into consulting. I traveled 80 percent of the time. I was gobbling it all up. I was meeting new people and learning how to think differently. It's what feeds my soul." She got married later in life, and her husband wanted to start having kids. "I didn't really know if I wanted to have kids because I liked what we had.

We dated for a really long time." Now that she has had two children, she tears up as she talks about trying to find balance. "It's making sure I'm a good mom and a good wife but also satisfying that need inside of me to always grow. And it's very hard because it's more than 50 percent on both sides. More than 50 percent of me wants to be a really good mom and a really good spouse, and more than 50 percent of me wants to work all the time. And that's really, really hard."

I ask her about whether this feels gender related to her. "I think it's who we are as people," she says, "but I also think it depends on who you marry. Whether you're in a heterosexual relationship or a same-sex relationship, it doesn't really matter. It matters how much your partner supports that part of you." She says this is true across sectors; she has girlfriends who "work all the time" in the financial sector and who "don't see their families that often." She married a more "old-fashioned person," and she thinks it's grounding for her, "because if he was like, 'Hey babe, go take that call that's going to take half of your Sunday,' I would, and I think it's good that he helps to rebalance me, although sometimes I wish I could just work."

Her method of finding better balance includes limiting work travel and having impact with each action versus working longer hours. "When you are trying to make it happen in the workplace and be effective and drive value with every action that you make, it requires you [to] have a more defined period of time in which you can do that. And when you're working in multiple time zones you also have to be hyperfocused; like, every sentence counts because you don't really want to have to do any rework. It means that you're taking away from other things that are also a high priority in your life." The prioritization carries through to how she thinks about spending her time. "I travel quarterly now and only for, like, four days. I exercise, and I use the words *want to* because it's a trigger, emotionally, for me. If I say, 'I should,' I feel like it's something that's being done to me versus something that I want to do. And tweaking that changes how I approach things in my life. I want to spend time with my kids, I want to do this, I want to do that. And it really helps reframe things for me mentally and emotionally." She received this advice from a former coworker, a CTO who heard her saying, "I have to." "And he goes, 'That shouldn't be a part of your vocabulary. It should be *you want to do these things*, and if there are things that you don't want to do, then you shouldn't do those things.'" I noted I'd heard a similar concept from another woman regarding "living a life of shoulds," and Frances said, "It doesn't feel good. It feels like you're living your life for other people. I made these decisions, and I made them sanely. So how do you make all of these things fit together? It's the norm for now. I'm just about to get to the point where my husband and I can go exercise without getting a babysitter. It's phases."

HAVING IT ALL

Don't talk to Bethanie about having it all. "I did a talk the other day, and the young woman who introduced me with all good intentions said, 'Bethanie's here to talk about career and breadwinning women and having it all.' And I was like, 'Oh no, let me stop you right there." In the last year Bethanie has started an internal company email alias for breadwinning women, women who are the sole or primary income earners in their household. The success of that effort has parlayed into other activities, like speaking at an event and hosting a podcast. But she doesn't like the perfection the phrase "having it all" implies. "First of all, I have *all of that. All of the things* are mine, but I don't have it all. At any given moment, you could ask anyone in my life if I'm letting them down and invariably, at least one person will say yes. There are times where I feel like a totally shitty wife, shitty mom, or times where I'd come back to work, and I'm not sure my [access] badge is going to work when I scan it."

Amid this imperfection, she values where she spends her time, going where she's needed most. In particular, she saw the opportunity to help breadwinning women and mothers. "We're a growing faction of society. Forty-one percent of American households are financially led by women. We are 50 percent of the population. We are actually graduating from college more than men. We are getting more advanced degrees than men." While she's not sure where this focus will take her, careerwise, "helping these women to find their community, identify the resources, identify their needs is really something I would love to make more and more a part of my day."

THE PRICE

Cathy chose her daughter over her career. Having taken years off to stay at home, she now does freelance work. For her, "it was important enough for me to spend time with my daughter in her early years. I'd be there for her whenever she needed me." The tradeoffs for her weren't as important. "For what? Yeah, I'd have more money in the bank. I might have a cushy retirement, I can buy a bigger house, but those things in the long run cannot replace the time that I missed with my child." While she may sometimes be stressed about money, it still doesn't make her want to sign up for a regular full-time job. In part this was due to a less family-friendly environment at her last company.

That said, she didn't want to admit to others she was staying at home. She ended up not working outside the home for five years due to a combination of issues that included health matters. "I did have a goal, as I was working on all my personal stuff, to return to the workforce. I really wanted to because, and this was a big struggle for me, during the time I was off for five years, I didn't tell a lot of people that I was a stay-at-home mom. I felt like if I said that, I'd feel like I'm one of those losers who doesn't have a job. And that's such a terrible thing to say because I love stay-at-home moms, and they're brilliant women. They chose to stay at home . . . but for some reason there's this internal, I guess, almost like a guilt inside me—why did I give up my career?" At the same time, she's really glad she took off the time for her health and her daughter. But still there was that tension: "During that whole time I was a stay-at-home mom, if they weren't a close friend, I'd be like, 'I do freelance UX work' . . . I had a hard time saying, 'I'm actually taking a break from work.'"

She thinks a lot about the parent tax we pay when we make that choice, in particular that mothers more often pay. "I feel very strongly that mothers are discriminated against. I think it's the overarching culture." She'd recently been reading a book called *All the Rage: Mothers, Fathers, and the Myth of Equal Partnership*, by Darcy Lockman, which talked about the societal impact of women continuing to bear the brunt of household and parental work. Cathy lists the penalties of being a mother she faces: the penalty of taking time off, onboarding again after time away, the need to leave work early for pick-ups, the ongoing interruptions a mom can face when a child is sick. "Even if the woman is really dedicated to her job, just the fact that you interrupt that work day and say, 'Oh, I have something else to attend to,' makes her seem as if she doesn't care about her job or that she's not ambitious, or her mind is so divided that she will never be a dedicated worker, or contributing as much to the company or the bottom line as a man." While there are men who do play a larger role at home, we're also seeing them face similar penalties, so this is a broader cultural issue to address for everyone, not only for women.

Beam also notices the inverse price if you stay career focused over all else. "I see the people who are directors and VPs in my company, and I feel like if I were in those positions that I wouldn't do anything but think about work. I don't think that I'd be able to compartmentalize work and family or anything else that I needed to do. I would worry about myself if I were in that position for only working and then one day dying and only being a careerwoman."

Did I mention there were challenges? Of course there are. Whether at work or home, I'm constantly rediscovering life's way of surprising me with new problems. But does that make me give up? Well, sometimes I need a nap! But no. After that nap, it's time to problem-solve and move forward.

In part 3, "The Power of the Example," I explore how women have navigated these challenges and built tools for building their careers. I'm also interested in who helped us along the way and what our tough days are like. Finally, do we want to stay in tech or are we planning to leave? And why?

PART THREE

THE POWER OF EXAMPLE

It's almost in our DNA as a woman to give back, that our work has purpose whether it's creating amazing tech to make the world a better place, creating an amazing team, or changing how well-being looks for employees. It's all about making a difference.

— Jennifer Lim

8 | Our Magic Toolbox

When I was twelve years old, I volunteered on the weekend with an animal rescue organization. They set up a booth in the local flea market, and we enticed passersby with cute puppies and cats. Volunteers walked and fed the dogs, but there was always an adult volunteer who ran the show. One of the adults did not like me. She was particular about processes in a way the others weren't, and I was raised with a healthy amount of independence, despite being a rule follower by nature. I believe I outright questioned her at various points. One day I got a letter in the mail: my services were no longer needed as a volunteer. My mom and I decided it was for the best. "They couldn't take your ideas" is my mom's summary, thirty years after this happened.

But. The dreaded but.

We all have to learn how to navigate the workplace and succeed. In future roles I learned how to share my ideas. I figured out how to meet a person where they were and how to further my goals versus hammer people with them. Sure, it would be great if everyone simply listened to me and accepted what I had to say. Even better if people accepted my authority without question. However, that's not what happens for most of us. In the 2018 McKinsey *Women in the Workplace* report, women continued to report a higher rate of microaggressions in the workplace, such as needing to provide more evidence of their competence than others, having their judgment questioned in their area of expertise, or being mistaken for a lower level than they were. This was especially true if they were the only woman in the room, with 51 percent of "only women" needing to prove their competence versus 24 percent of women who were with other women and 20 percent of "only men."[36]

---

36 Krivkovich, Alexis and Marie-Claude Nadeau, Kelsey Robinson, Nicole Robinson, Irina Starikova, and Lareina Yee. 2018. "Women in the Workplace 2018." McKinsey, October 23, 2018. https://www.mckinsey.com/featured-insights/gender-equality/women-in-the-workplace-2018.

For me, a positive attitude and a huge trunk of skills were required to keep going in the business world. These aren't the lessons I learned in school, but the ones I learned by doing. As I talked with women, I asked them for a key challenge they faced in their careers. We discussed how they approached those situations and what tools they developed. Women spoke about what they had to learn about other people and the need to market themselves. They also learned about themselves: their strengths, when to listen to their inner voice, and when to move on. Taking control of their lives was also a key theme: how to ask for help, get support, and set boundaries. And the traits they needed to build: confidence, grit, and perseverance.

In the end, I've boiled down everything I heard into a top five list of tools for navigating a career in tech, and other places as well. The world of career advice is cluttered with all types of recommendations, so I'm going to clean this up as much as possible and tell you what I think was most effective for myself and the women I interviewed. My recommendations are based on not only how often I heard the advice during interviews, but also the relative impact of different types of tools (i.e., which ones made a big difference in how we felt and what we did). In addition, none of this advice is groundbreaking. I'm standing on the shoulders of experts like Angela Duckworth (*Grit: The Power of Passion and Perseverance*). If you are already employing some of these tactics, that's amazing! It's my hope our stories and experiences help bolster that work you are doing, remind you why you are doing it, and aid you to help others. I will also refer to other resources for additional research and learning. If you are just starting out, I hope you can use this as a touchstone when you aren't sure what to do next.

The top five tools covered in this chapter are:

- *Resilience*: building the grit and the power to withstand adversity
- *Marketing 101*: speaking up and promoting our talents and accomplishments
- *Ask!*: having the confidence to network, reach out, and ask for help
- *Find Support*: seeking out those who will help us navigate our careers and provide a sense of belonging
- *Own Your Awesome*: Knowing you are enough and you are worthy

RESILIENCE

As I spoke with women, resilience as a powerful tool came up over and over again. They may have called it different things, but it was all the same notion. Some called it thick skin, bravery, grit, perseverance, being tough, or even being oblivious, but it all boiled down to the same skill. As Angela Duckworth describes in her TED Talk, "The Power of Passion and Perseverance,"[37] "Grit is passion and perseverance for very long-term goals. Grit is having stamina. Grit is sticking with your future, day in, day out, not just for the month, not just for the week, but for years, and working really hard to make that future a reality." That is what all the women were speaking of: the ability to last, the ability to succeed despite struggle, not because the path was easy. Or even better, the ability to succeed *because* of struggle.

"I wish I'd taken more psychology courses in preparation for being a manager and leader," I said to a fellow director recently, echoing what I've said many times in the past. We were puzzled over a common situation—how do you figure out why a highly talented, highly motivated person is stumbling? Even at a place like Google, filled to the brim with overachievers, we regularly see people struggling to succeed. Not all of them are familiar with that challenge; many have been successful at everything they've done up until that point. Helping people develop and grow their careers is one thing, but nurturing people's emotions, self-acceptance, and perseverance is a whole other skillset.

I chose to call this section "Resilience" because, per the *Oxford English Dictionary*, it means "the capacity to recover quickly from difficulties." That is a skill you can build, a muscle you can strengthen, versus something you were born with or a certain background you need to have. In recent years, the business world and schools have embraced the idea of a growth mindset developed by Carol Dweck at Stanford University. As Angela describes it, "it is the belief that the ability to learn is not fixed,

37 Duckworth, Angela Lee. 2013. "Grit: The Power of Passion and Perseverance." Filmed April 2013 at TED Talks Education. Video, 6:01. https://www.ted.com/talks/angela_lee_duckworth_grit_the_power_of_passion_and_perseverance/.

that it can change with your effort. Dr. Dweck has shown that when kids read and learn about the brain and how it changes and grows in response to challenge, they're much more likely to persevere when they fail, because they don't believe that failure is a permanent condition." By teaching ourselves to learn from the hardships we face, we can build resilience and use those tough days to fuel our great ones.

PERSISTENCE

Ashley doesn't sugarcoat her college experience: it was difficult to follow through with her EECS major. "It was very hard for me. I had moments in finals where I was like, *I'm gonna fail this class. I need to start replanning my schedule for next semester because I have to retake this course.*" Ashley describes how some of the introductory coding courses in college had close to a thousand students, requiring overflow rooms nowadays. With classes so large, she noted, "it was easy to feel lost and really stupid when you don't understand it. You're scared to ask for help." Even when she went to office hours, she found other students there who were only going to help fellow students, not because they needed office hours themselves. "I thought the community was very cool, but I think it was in college I then started getting that idea that I was dumber than everyone else." Despite these challenges, she worked hard, persisted, and made it through, obtaining her first job after an internship.

By sticking it through, she feels like good luck then came her way. She was the first developer on the developer operations and infrastructure team, so early it wasn't even called that yet. The company was also growing rapidly, had gone public, and was doing a major migration from a data center to hosted service (AWS). She was thrown into responsibility, with only Ashley and her manager on the team. "I think as a fresh, fresh grad out of college, you normally wouldn't have that big of a project." Since she was the only one there, however, "everything kind of fell on me, and at the time I didn't know it was such a big responsibility, but looking back, I would not have trusted my own self back then. I got to build all the deployment software, first in the data center and then all the software that we used to migrate from the data center into AWS." While I mostly attribute this opportunity to a ton of hard work, Ashley also sees the fortune she had. "I always say I'm lucky because the timing was good, the team was good, my manager was really amazing. He always trusted me and gave me these big projects, and he was always promoting me to other teams and higher-ups."

Alex dealt with the insecurity of starting "late" as compared to her peers. "Even though I started coding at a fairly young age, around seventeen, it still felt very late compared to my peers. My first years in college were the hardest, as I would hear classmates rant about the latest technologies and have conversations using technical jargon every three seconds." It wasn't a welcome environment, and not always as it appeared. "It took a while for me to understand who is being authentic and who is just trying to intimidate. At the start, the intimidation was very harsh, and I continuously questioned my abilities as a developer." Within this challenge, she found a passion for the work and the grit to continue. "I didn't like the feeling of being 'below average' or even 'stupid' at times, as I hadn't experienced that prior to college. I took it as a challenge and decided to work even harder to catch up. Somewhere along the way, I stopped doing it as a competition and just worked hard because I liked the work."

Through this, Alex has begun to figure out ways to navigate the industry and remain confident, especially in a fast-paced work environment. "The work is rapidly changing, and one must learn new things at all times. There's no time to get bored, and for the most part the community is really eager to help. Asking questions and collaborations are often very encouraged, at least on paper." Despite this positive message, she does warn that women may encounter some additional headwinds, citing examples where it's still hard to convince men that you're equals. "I've had experiences where it takes the other parties a while to realize that I am just as experienced or knowledgeable about the matter, and my ideas might even be better than theirs on some aspects. It's impossible to read people's minds and determine if it was circumstantial or there's a gender bias, but I'm cautious of this." She doesn't want to alarm others, recommending awareness and sensitivity to the situation. She tries different communication methods when needed, and regardless, reminds herself "that my ideas are valuable and deserve to be discussed."

Sarah has also seen the benefit of being thrown into the deep end and learning how to swim. "When I first joined, I had never taken a single business class in my life and was thrown right into senior leadership meetings where my role was [supposed to be] to take notes, interpret what was happening, then translate it all into executive communications and presentations." Despite being tough, this experience taught her more than any training program. "I always consider this my 'paid MBA' because I learned *so much* from getting that level of access and visibility to how executives interact and make decisions." Like others, she felt insecure at first: "I always felt like I was faking it, like I was severely underqualified, and that sooner or later people would realize I had no idea what I was doing. But I kept going. To my surprise, everyone loved my work. I even had senior leaders requesting my help in helping them revise their resumes in confidence."

Mai talks about the everyday wear and tear to persist. "I think that, looking back on my own career, I've tolerated a lot of jokes or a lot of side comments or mansplaining." She remembers when someone tried to explain Slack to her as if she didn't already know how to use it. "Those moments really do ruin your day, because it's not about the work that you're doing and the stress of the deadline. It's actually about the interactions with your coworkers that sometimes don't . . . feel that they're doing service to you, that they ended up wearing you down and making you feel like, *Duh, dude, I've been in tech longer than you have. I got it.*" This story reminds me that persistence, alone as a tool, has a price, and we do have to watch out for how it wears us down or affects how we feel at work.

FAIR OR NOT

Camie started as a software engineer in the late eighties. She joined a team where men were in the majority but that also had 30 percent women. For the first nine years of her career, "I either had a woman manager or a woman second-level manager or both. There was always a woman above me." She then left to start her own company and had a very different experience. "I paid no attention to [the fact] that I was black. I paid no attention to [the fact] that I was a woman because I had been in that environment. I realize now I wasn't aware or thinking about diversity upfront, and I was just happy to have anybody that would join and work for me." She attempted to recruit women from her network, but ultimately she was the only woman on her founding team. "I was so busy trying to make the thing successful. I wasn't even really paying attention to it. And then I think over the years I realized that I was the only woman in the room, but it didn't really dawn on me that it was like that in *every* room. It wasn't until I read *Lean In* and started seeing the data, I'm like, 'Oh shit, this is bad.' I was heads down focusing on what I was trying to do and not paying that much attention."

For better or worse, she notes how she took full responsibility for her success, never assuming that anything had to do with her being a woman or black. "I always assumed wherever I hit hurdles, it was because of me innately. And so it was up to me to figure out how I could be more successful." She spoke about having "Teflon skin," the ability to move past, ignore, or not even notice slights or microaggressions. While she understands why some actions can be viewed as offensive (e.g., the overflagging of certain people for certain types of work), she's also noticed that many women who've done well have simply moved past and succeeded despite these issues. "The microaggressions don't get to them. They either ignore them [or] they don't let them get very deep. I think that's how you stay in a positive mindset, because if you get into a negative mindset it doesn't help anything."

This struck me as both true and tricky. What happens if women become frustrated trying to resolve blocks or barriers all on their own? What happens to the women without Teflon skin, who might be just as or more talented? Camie remembers fielding questions from a black woman about working in tech. "When I told her what I did to survive, she asked me, 'Isn't it hard?' And I thought, *Well, I never expected it to be easy*. And she asked me, 'Do you think it's fair?'" In response Camie said, "Well, whoever said the world was fair?" Her response was ingrained by years of challenges. "I learned from a young age: the world was not fair. Rules didn't apply to me the same way they apply to everybody or other people." She didn't grow up with "the expectation that success was going to be easy," so she was prepared for challenges in her career, while others might be less so, depending on what they expected.

Michelle flags the survival bias implicit in expecting women to have thick skin or favoring those who do. She remembers an activity a few years ago "where my senior VP at the time got a whole bunch of senior women in his org together to do a listening lunch or something like that. There was all of, I don't know, nine of us. We were gathered together for the first time, and I looked around the room and I suddenly realized, "*I have a lot in common with these people.*" She realized all the women fell into a similar bucket. "We're all, 'I like video games. I would love going to drink beer after work with all of you. I've got a knife in my boot, like, fight me.'" (That last one was a joke, for clarity's sake.) She observed that this wasn't a typical group; it was a senior group of women who had survived earlier challenges. "I absolutely in that moment was like, *Ah, I feel like a caricature when I look around this room. This is not common.*" Similarly, Rachel looks back and sees a pattern. "I think a lot about the women that I know who are successful, and one of the things that most of us have in common is that we don't take any shit." She wonders whether that's a selection bias, "if we would have been selected out if we weren't like that. If we were soft spoken, would we have made it?" She feels like it's almost a requirement in tech, but she also knows that she learned how to cope over time and is now someone who will stand up for herself and others. "I think when I first started, the feedback was that I had to speak up more. Clearly I conquered that."

BRAVERY

As a director of engineering, Michelle speaks of working on being brave beyond the basic comfort of speaking up. "I've gotten good at having hard conversations, so it's not that style of bravery, and I've gotten mediocre—like, good enough, I would say—at mentioning elephants in the room. So it's not quite that either. It's more creating conflict that needs to happen." She'd like to be able to provoke the right kinds of conversations to drive action forward, "where it's like, oh man, if we just had a fight right now, the conflict would be over." She's looking for opportunities to try it out.

Jill explains why speaking up might be necessary in a world where the best idea wins. "There's an engineering way of thinking, which is arrogant with a small *a*, like 'I can do this better.' Women have that too, and many women engineers have that mentality too. But there's a model of engagement with often-male engineers that is not for everybody, because you do have to fight your way to win the argument." She can't imagine that changing, so her advice is to prepare. "For folks who don't enjoy [that] or find that kind of engagement really off-putting or upsetting or triggering, you have to prepare yourself for it." While individual engineers' personalities will certainly vary, it's worth noting that the debating of ideas and opinions is a common element in tech so you don't think you're alone or believe it reflects on you personally.

BEING KICKASS

Many times, Hillary sees that people may not even be aware of the bias they bring to the conversation. "I was talking to a very senior leader, and I was doing my thing in one of my meetings. He was like, 'Wait, you're a mom of three. Don't your kids need a mom at home?' I called him on it and said, 'They don't need me at home, but they need me and they need their dad, and we're all there for them. Thank you so much for your concern.' I think he didn't even realize what was coming out of his mouth and how incredibly backwards that was. But I get that kind of stuff daily." She finds it interesting what people will say to women that they'd never say to men. "People are more aware of it and behavior has changed to some degree, but I think it's still there." To people who have concerns about being a working mom of three, she says, "Will I be able to get my job done? Are you serious? If you want something done, give it to a really busy mom. She will make it happen, and she will clear every obstacle to get the job done because she doesn't have time to mess around."

Amid possible adversity, Hillary's advice is to be unshakeable. "Be really confident about your education and your experience, and do great things and make that speak for what you're going to achieve, because that's the way that change will happen." She recommends women be "kick-ass at what they do and make sure that nothing will stop them that they can't control. They can control their performance and where they're going to take their career." And it's up to you whether you speak up. Hillary's view for herself is, "I think for all womankind, and for mankind, generally speaking, I need to say something. I can't let things go without acknowledging [them], but I don't let that keep me from doing my job."

RETAINING HOPE

With the need to persist, people can lose hope, get tired, or feel the world is against them. While that's natural in some ways, I've also seen how successful people are able to, time and time again, resist the darkness associated with disappointment and frustration. Staying hopeful, despite hardship, is an incredible skill and strength. This can be particularly true when we're trying to drive societal change, which can take a long time. As previously noted, tech is like the rest of the world and is still midchange when it comes to biases and diversity. Frances encourages us to, "before making assumptions about why things are the way they are, seek to understand why things are the way they are and try to influence how they can be better." In the last five years, she feels "like people have lost a sense of assuming positive intent, and they assume that things are the way they are from a negative perspective, which is not the right way to go." She notes the downsides of that position: "It takes a lot of energy to think that way, and you're not going to make any friends. You're not going to be able to influence anybody, and you're shut down. You're not open to hearing things a different way, and you're not going to help other people be open to how you are thinking either." She encourages everyone, men included and across industries, to "be open and assume positive intent and help people understand how you're feeling, and be open to understand[ing] why things exist with the understanding that things can absolutely change. They might not change as quickly as you would like them to. But look for those incremental changes that can happen and look for those bargains along the way."

As a multiethnic woman, she's seen firsthand how companies may change slowly because they are often first focused on survival. "When you're at a company where the focus is growth, the focus is growth. It's not on developing people. I mean, unless you're at a big company, like five thousand–plus people, usually you're at bone and hard muscle and you just don't have the resources to do those important ancillary jobs." She notes this may make a company look less diverse than it is. "When you have a company that's in hypergrowth mode, maybe they're not looking to update their website, but it isn't because they don't want to." She's also seen how the move toward diversity and inclusivity training has brought perspective to the conversations. "People are able to share what some of their life experiences were, and interestingly enough, we all have so much more in common than we don't."

Paola has a similar feeling. "You can't get yourself in that mindset of *I don't see [someone like] me, therefore I am not good enough.*" She doesn't let that stop her. "I just think, *Okay, [these] are areas of opportunities that I can do to bring more people like me in the pipeline.*" At conferences or other industry meetings, she'll watch out for other Latina women. "I reach out to them. I have them apply, and then I'm there to support them on their questions." In particular, she wants to be there because she was in the same place once. "I felt that imposter syndrome . . . that question is like, *Am I qualified to do this*? And so I support them in that [by saying,] 'Hey, this is my story of how I even got this role. And how I got to where I am.'"

A quick note on imposter syndrome, because it will come up a number of times in this chapter: simply defined, imposter syndrome is persistent self-doubt. The dictionary version presents a double bind: "the persistent inability to believe that one's success is deserved or has been legitimately achieved as a result of one's own efforts or skills." This means that even a successful person can suffer from imposter syndrome. In fact, they may be more likely to because they will have achieved success and been recognized for it but not *feel* successful despite all their work to get there. The quotation I always use to typify imposter syndrome is "one day they'll find me out"; they'll figure out I'm a fraud and not really that smart, talented, and so on. Sarah illustrated how imposter syndrome has shown up in her career via examples:

- Passing up or not going after various development opportunities (training, conferences, mentorships, stretch projects) out of fear that I would be perceived to be overambitious, asking for too much.

- Not applying for roles out of fear that others wouldn't feel I was qualified, or worse, out of fear that it would be perceived that I was not thankful for the support I have in my current role. (The subcontext here is that I am fortunate to have support and should prioritize being grateful, minimizing the fact that I earn the support each day through hard work and results.) I think that is deeply rooted in coming from a different background, not meeting the on-page requirements for this job, at least initially, and feeling like folks took a gamble in hiring me and I should give some loyalty in return.

- Waiting for validation from a manager, coach, or mentor before pursuing next-level things.

- Second-guessing decisions to the point that it causes procrastination and delays outcome/results.

Both men and women can face imposter syndrome,[38] but coupled with other issues women may face as minorities, this comes up frequently as a significant issue for women as they persevere and navigate their careers.

LEARNING FROM THE PAST

Jill built her resilience early in her career, having founded a start-up that "did not make it through the success barrier." The start-up was able to raise money and had built a team of about twenty people, including engineering, product, and sales. The company floundered due to its sales tactics and strategy, however. Amid the struggle, their office was destroyed on

38 Bravata, D.M., Watts, S.A., Keefer, A.L. et al. Prevalence, Predictors, and Treatment of Impostor Syndrome: a Systematic Review. *J GEN INTERN MED* 35, 1252–1275 (2020). https://doi.org/10.1007/s11606-019-05364-1.

September 11. "That's a big setback, obviously, for many reasons: psychologically, physically, momentum, the world's banking, all that stuff. But we rallied through that and tried to overcome that huge obstacle in our path." Despite their efforts, she had to shutter the company, which impacted her deeply. "I felt this deep responsibility for all the folks who had left other jobs and signed up to work for this company and were really inspired by what we're trying to do and by their colleagues and how we were solving this problem for people. Having to shut down the company and figuring out how to take that failure and figure out how to productively learn from it was a huge challenge." She describes how, being "a type A overachiever who had done well in school and then gotten into a great college," this was her first significant failure. She lists the difficulties involved in figuring out how to learn from her experience, how to "absorb the lessons without letting it set me back from a confidence perspective, knowing what pieces I should take and try to grow from personally versus knowing which pieces were, hey, I made a poor business decision here, and I need to learn from that." At the same time, she was helping her employees with connections so the impact of the shutdown was minimized. That period was extremely difficult, especially after 9/11, but it taught Jill important lessons for the future: "What it certainly taught me was that it's possible to get through anything, and it's really, really important to bring people along when you're enlisting them on a journey. You owe it to them to bring them along every step of the way."

Gretchen, now a COO, echoes learning from hardship. "Sometimes you can learn the most about who you want to be from the worst leaders. During my first job out of college, a senior manager publicly berated me for his own technical mistake in front of a client executive." She was shocked and discouraged to see him blatantly lie, and as an entry-level employee, she "was too intimidated to address the situation directly." It stuck with her, though, and the incident "taught me a lot about respect in the workplace and helped inform the type of leader I wanted to be."

RESOURCES

Where else can we learn from? While doing research for this book, I noted down resources I thought were particularly valuable. These resources, mainly books, will go deeper into each of the areas, and they will also provide more tools to dig into if you'd like more information. I also list them, along with other resources and perspectives, on my website at alanakaren.com.

Grit: The Passion of Power and Perseverance *by Angela Duckworth*

I saw the video clip from Duckworth's TED Talk, and I was all in. Her work is all about how we succeed through passion and perseverance rather than innate talent. As someone who works very hard but does not consider myself *that* smart, she is speaking my language. While some criticize the book for being repetitive, I enjoyed the stories and narrative paired with the research of why grit matters.

Mindset: The New Psychology of Success *by Carol S. Dweck*

An often-referenced resource both at work and at my children's schools, Carol Dweck's work has paved the way for a new understanding in how we succeed. Dweck is referenced multiple times by Angela Duckworth, in fact. As a history major, I always like to go back to the source material. I was surprised at the breadth of examples covered in the book across sports, relationships, business, parenting and teaching. A worthwhile scan even if you think you know what Dweck is all about.

The Mental Toughness Handbook: A Step-By-Step Guide to Facing Life's Challenges, Managing Negative Emotions, and Overcoming Adversity with Courage and Poise *by Damon Zahariades*

Zahariades differentiates mental toughness from grit, with the former being a state of mind versus grit as an attribute, an inclination. I didn't pause much to debate that distinction, being drawn in by Zahariades's clear and succinct writing as well as abundant, practical exercises (short ones indicating time required for each) to train our minds to be tougher and stronger.

MARKETING 101

I have a weird relationship with trying to promote myself. I actually think I'm pretty awesome, but I also know that I'm not the hottest thing around. I consider myself smart but not the smartest. Cool but not the coolest. I wouldn't even call myself modest, more truthful with a cynical bent. That is, until I started to write my book.

It turns out when you are trying to get a book published, the road is much easier if you are the most famous, smart, cool, hot thing around. Publishers are interested in the platform you've built—how many followers you have, how many conferences you've spoken at, and how many reporters have covered you. But oops—while I was busy solving problems internally at Google and having three children, I hadn't invested in my Instagram fame or prioritized speaking externally. In the official ways things are measured, I wasn't awesome. Yet.

But as I started to figure out how to build up my external reputation via LinkedIn and other means, I was still holding myself back. As part of a company training that covered why it's difficult for people to change,[39] I realized that being humble is such a strong value for me that I'll hold myself back, even from goals I strongly desire, if I feel my humility is at risk. I instinctively dismiss certain types of social media posts I deem purely promotional versus valuable, even though they are extremely common. I'll play down my experience in bios or when describing myself, to the point others will correct me. How do you market yourself without . . . well, marketing?

I thought a lot about what to name this section. All the terms—*self-promotion, self-marketing, branding*—have negative connotations, especially when it comes to women. Per the introduction to the "You Have to Be You" chapter, women are expected to be humble and can face penalties when thought to be bragging. Nonetheless, we ladies are going to have to play offense if we want to get in the game. One of the best ways?

---

39 Harvard Extension School Blog. 2019. "The Surprising Reason We Don't Keep Our Resolutions (and How to Overcome It.)" Published January 7, 2019. https://www.extension.harvard.edu/inside-extension/surprising-reason-we-dont-keep-our-resolutions-how-overcome-it.

Get very good at and very comfortable talking about yourself. As usual, I turned to the women I was interviewing not only for their stories but also for their advice.

WHO I AM

As I spoke with Amy, I mentioned that she was amazing at building and promoting her resume. I've seen her posts on social media about recent events she's hosted and attended, for example. She's being proactive in that area, and I asked whether she's found that helps her career. "It seems self-serving sometimes, but I've been thinking a lot about my brand and who I am and what I project out onto the world." She wants to project an image of competency, engagement, and commitment to community, leadership, and diversity and inclusion. She is working to specifically tailor her "outward branding" toward those areas. "You do have to tell people what you bring to the table." She has seen people get laid off or not get a promotion, and she knows it's not just about the work that you do but also how you sell yourself. She has found that when she does that, unexpected opportunities arise. Her current job came from Tieisha, also interviewed for this book, whom she met in 1999. "So eighteen years later, because I'm conscious and I stay in touch with people, they're like, 'I know who you are, you would be great for this.' Those seeds flower in very unexpected ways, and I am very conscious about it."

Carol also learned that sharing more of who she is was important to making an impression. "When I was young, or when I just joined the workforce, and whenever I joined a new company, I tended to be a little shy." She focused on her job and thought that would be noticed. "Later on they will see the talent or how capable I am; everything will be good." Over time she realized "impression and perception and having that presence is very important," and she recommends "not to be afraid to demonstrate who you are, even during your first encounter."

SPEAKING UP

Sharing our thoughts and knowledge is a version of self-promotion, but it can be hard at first to know how to intervene. Diane describes an example: "I was in a room with two directors and [a senior manager], and they were talking about something I was working on." She noticed that the people were positioned so that the person at the whiteboard "kept looking between the two men in the room and not [at] me, even though it was my project." Now she'll speak up instead of thinking "this is how things go." It took time for her to become aware and feel like she could "just start talking" though.

Similarly, Marily says, "It took me a while to realize that being shy would never get me anywhere. I would miss out on opportunities by not raising my hand at school or by not speaking up during meetings." While she thought "my turn will come," she realized over time that "[I] needed to speak up for myself, express my opinion, and advocate for my work." When she did so, she saw her responsibility and credit increase, and "it felt great!"

She thinks this took some time because she spent years being the only woman in a classroom full of men, and she felt that "I earned my spot by luck, without deserving it." As she learned about imposter syndrome, she "realized that I was actually great at what I was doing, my grades were very good, and so was my code. I was shaping up to be a great computer scientist, and I could see that on a daily basis. I do feel that part of being shy was largely because I was the minority in the room, and I worked with myself on not letting this affect me."

Karen remembers thinking that "[management] will recognize that I'm already an experienced manager and they'll move me up." That wasn't happening, though, and a colleague told her she had to be entrepreneurial. While she wasn't used to thinking that way, she does think that her age, fifty at the time, gave her credibility. "Nobody ever [asked], 'Does she know enough about tech?' I knew people. I was already established." That gave her the confidence to assert herself. "At some point I realized, I'm going to have to butt in and speak up. And that's being entrepreneurial in a certain way. Then I began to think of things that I could make happen or I could do on my own, which led me then to manage the whole Google blog process, which led me to say, 'Google has to be on Twitter and here's how we're going to do it.'"

While speaking up was difficult at first because it was unfamiliar, she grew more confident as she got to know the company. "I didn't have that confidence until I was really familiar with the place and felt like I was part of the furniture." She also credits her ability to make connections and market herself to her innate curiosity and the ongoing exposure to different teams and interesting topics. "I was more curious, I think, than a lot of people. I wasn't just there for my boxed-in topic because . . . I viewed [my job] as an editorial person: Where are the stories? What are the themes across the company? I was in a fantastic position to see those. So that led me to be in touch with more people and meet more people across the company." She recommends curiosity and getting to know people as a long-term tool, not only a practical short-term need. The people she met in those years have led to sustained relationships that have aided her career over the years.

LEARNING TO PIVOT

Camille has felt a particular need to market herself as she attempted pivots in her career. It was at those times that she had to sell her experience as being translatable to other fields. Once she had worked for three years in the corporate childcare space, she felt a hunger for a new challenge. Then the question was this: "How do I pivot? How do I market myself for something else?" In a growing tech company, she theoretically should be able to find another role. "But then when you come in with this special specialty that is not common at every tech company, or even in the company that you're in, explaining to people where you can add value was really difficult."

She transferred to a more general people operations and human resources space while she figured out where to go next. What worked was "very aggressively interviewing and looking and pushing myself out there" until she found a role that had "the clear connection." Finding a role with clear ties to her previous role helped her pitch. "It was such a bright light that there was this clear connection: the amount of chaos, four different teams in four different cities doing things four different ways, which is what I walked into for different childcare centers doing things four different ways, and everything finally click[ed]. That got me into the mindset of, oh, these are the business skills that are needed no matter what [and] that I have that can be applied to almost anything."

LEVELING UP

A cousin to marketing, an ability to present ourselves and appear polished is also expected, especially as we progress in our careers. Camie described the skills she needed to build once she started her own company after being a hands-on software engineer for the first nine years of her career. "That was a huge challenge for me, because I had avoided any management responsibilities, and I don't think I had even given a presentation before." As a full-time coder, she felt the pressure of onboarding professional skills quickly. "Leadership skills, presentation skills, management skills—this whole slew of [1:1 and organization] management in a really short, pressured time." It was critical she onboard this new tool chest because the venture capital firm that funded her company had wanted to bring in an experienced VP of engineering, and Camie refused. "And so they said, 'Well then, you've got to get trained.'" Examples of words she uses to describe this period are *metamorphosis* and *painful*. The VC brought in executive coaches to help her, including a presentation coach and a "neurolinguistic coach that helped me with my speech patterns and presence."

She thinks this was a necessary learning curve. "I had to stand in front of investors and convince them to continue to give us money. We had partnerships, so I had to meet with technical leaders of other organizations and convince them to partner with us. I was a founder of a company, so I needed to inspire not just my team, which I had really always worked with one-on-one, but this company grew to 150 people in the four years that I did it. I had to stand up in front of the entire company. I'd never given a presentation before." She also wasn't the type to focus on these skills one-on-one either. "In those early days, I walked into our office and walked by everybody, sat down at my desk, and started coding. My founder sat with me and said, 'Hey look, you're a founder. You have to come in and say hello to people when you walk into the office. You can't just walk by them.'"

Similarly, Christine has felt the need to step up her game as she's become a senior leader, despite being a natural introvert. "I would say my personality is one that's more of someone who likes to be backstage, do it from the background. But at the same time being a leader of a team, one of the [pieces of] feedback that I've gotten in the past from [peers and team members] is wanting to see me more in the front." That's an area she's pushing herself to get more comfortable in, whether it's speaking up at executive meetings or in front of large crowds. "I've never had a problem speaking in front of my own team regardless of size. It's more when I am onstage or when I am talking to a big room of people that I'm not familiar with."

For her, practice makes perfect. She welcomes new people to the company on a monthly basis, using the same slides as a way to refine the presentation and build comfort. "I feel like part of it is doing it more, pushing yourself to do it." Another thought has helped: "I think one thing that has been really helpful to me that I realized seven years ago is managers are not perfect." She was able to see her previous leaders model being strong but not perfect. "It's okay that they make mistakes and they share; they were transparent about it, and they own it. Just seeing who they are and their endearing personality." She feels like this dispelled a need to market perfection. "It's less stressful, right? Being in front of the room. If you've messed up, then it's okay. You know there'll be another chance, and no one is perfect."

RESOURCES

For this section I wanted to avoid the whole "how do I become an influencer" vibe since that may only apply to some of you. Based on my personal experience with a broader concept of self-marketing, here's what I suggest. Your results and tastes may vary.

Just Do You: Authenticity, Leadership, and Your Personal Brand *by Lisa King*
Geared toward current and future leaders but applicable more broadly, this book goes deep into analyzing and understanding what our motivations

are and how we are perceived by others. There's a companion workbook available online with numerous exercises that help break this down.

You Are a Badass: How to Stop Doubting Your Greatness and Start Living an Awesome Life *by Jen Sincero*
This one is written like a close friend is cheering you on. I liked the informal, down-to-earth narrative, and part 2, "How to Embrace Your Inner Badass," made some good points. For example, how I should stop my habit of self-deprecating humor as it's self-destructive and also hurts how others perceive me.

On Instagram, I also suggest @thealisonshow, an influencer and branding guru with a podcast, blog, and branding school. Her posts are big on how to be kind to ourselves as well.

ASK!

I was recently participating in a panel at a tech company. The topic of discussion was women in tech, and I answered the question, "What one piece of advice would I give my younger self?" I said, "Just ask." Ask for what you want, ask for what you need, ask, because the worst thing is to look back and realize you should have asked.

How I grew up has left me with a permanent hangover of complacency. I tend to assume something won't work or isn't possible. Even worse, I won't ask questions or think about what could happen. This has applied all over my life, but probably nowhere worse than at work. Years ago I was wondering why my manager wasn't offering me additional responsibility when I was performing so well in a related area. I didn't realize that my coworkers were asking for the opportunities they were receiving. A lot of life has been a mystery that way.

And it turns out other women face the same personal obstacles. A 2019 LinkedIn Gender Insights Report describes how women and men find jobs differently. From their data, women are 16 percent less likely than men to apply to a job after viewing it. Women also apply to 20 percent fewer jobs than men. As noted in other research, women feel they need to meet 100 percent of the job criteria while men typically apply after meeting about 60 percent. Women are also less likely to ask for a referral from somebody they know at the company.[40] This is all a form of asking, whether for a job or simply for help.

In the last few years I've been trying to break this pattern by asking myself two questions:

Why not? If I think of something awesome I could do, I stop the nay-sayers in my head and push forward with the simple question, "Why not?" When I saw an application for a customer service award for my Google Fiber team, the deadline was days away. But I said to myself, "Why not?" Months later, we were going home with a gold award.

What would rich people do? Seriously, folks, this works for me. Sample conversation with myself:

— Self: Should I write a book?

— Self back: Why not?

— Self: Okay, but how should I get started?

— Self back: Huh, what would rich people do?

— Self: I guess they would find a person who knows what to do, and they would get advice.

— Self back: How would they know how to find that person?

— Self: They would know someone? [pause]

— Self: Do you know someone?

— Self back: Wait, I do. I DO KNOW SOMEONE.

Always ask for help. It's not a sign of weakness. It's not just something other people do. It's the best thing to do. In the last year, I've reached out to various people I know directly and others to ask for publishing advice. Has anyone turned me down? No. Have I received invaluable feedback and support? Absolutely.

It can feel ridiculous to go through a conversation with yourself, but I know I'm retraining my brain. I'm pushing myself to see possibilities, to ask for help and try new things. If you are a manager, I hope you'll keep

---

40 Ignatova, Maria. 2019. "New Report: Women Apply to Fewer Jobs Than Men, But Are More Likely to Get Hired." LinkedIn Talent Blog, March 5, 2019. https://business.linkedin.com/talent-solutions/blog/diversity/2019/how-women-find-jobs-gender-report.

this in mind as you look at your employees. Perhaps the ones who don't seem ambitious could simply use some help to see what's possible. Here are other stories about the challenge of asking, and how women grew their skills in this area.

ACCESS

Michelle feels like a recurring challenge in her career is getting access to the opportunity. "It feels like I have to fight to just be given the chance for things." She feels like once she has the work, then doing a good job "is the easy part." It's not always new jobs or new tasks. "Sometimes that comes in the form of access to rooms, access to participating in decision-making processes; sometimes it comes in the form of being considered for a thing." She's had to shift her thinking "to think about my own PR in a way that'll allow me to get opportunities."

She did notice this got better as she became more senior. "One thing that changed, quite frankly, that made my life a million times easier is I got promoted to director. Having that title helped way more than I feel comfortable that it should have. Walking into meetings, meeting someone for the first time, there was an assumption of competence that really fucking helped across the board." To get to that point, though, "I tried a billion different things. None of them big, but some of which helped a lot." She gave examples like "being cognizant of finding ways to make sure my achievements were known without gloating about them. Because if I talk about my achievements that's bad, if I don't talk about them at all, that's also bad."

The most common advice she gives to junior people? "Ask for what you want." She'll have people come to her and say they were interested in a project but it was given to someone else. "I'll be like, 'Did you tell anybody? Are we expecting them to be psychic?'" Another example is where there's conflict and she's told, "'I'm having this conflict with this other person and they're doing A and I wish they were doing B.' And I'm like, 'Have you talked to them?'" As a minority in tech, she shares, "one of the things that comes with being less listened to and catered to by default is that people are not going to notice if you're not happy. You have to be even more overt and explicit about those things." Her overall take is that things get better when people ask for what they want.

BIT BY BIT

Kris gave a reminder that we don't all of a sudden have to change how we behave. "It doesn't have to be a light switch. It can be incremental. Just ask a question in a meeting. Yes, you're the most junior person and you're the only woman in the room. That's okay. Start small." She wants to make sure we're mentoring, coaching each other, and actively resisting "this idea that you have to either be that way or not that way." Her experience is "that all women are capable of being more in the room. You just have to have the right strategies to get there."

One of the ways she's asked for more is by educating the men around her to help make them advocates. When she was in tech support, her entire team was men, and she would share stories of what she encountered. "I was the manager of the whole team, and I was taking a call from someone and he said, 'Can I talk to a man, another person? Can I talk to your boss?' I'm like, 'Well, my boss is the VP. So you're talking to the boss.'" The men on her team were horrified and couldn't imagine this was still happening. She's learned that "educating the people around you of the realities that exist for women is incredibly important, because I think if you don't talk about it, then other people don't talk about it. I bet you five dollars at least one of those men on my team went and told that story to somebody else with that sort of incredulousness."

Peipei reinforces something we can do slowly over time by not taking on helper tasks like planning parties or events. "Ask for help. Know it's not our responsibility to do those things, but unfortunately this is what helps you set your own boundaries." If it's not your job, ask for help from other people. "If you're experiencing things that are uncomfortable, just ask for help from a person that is going to support you." Most important, focus on "developing your domain or technical depth. I think that it's great to do other things, but this is not the woman's responsibility. Don't always be the person that volunteers to plan the off-sites or birthday parties and baby showers." She reminds us that society may look to us or expect us to do these things, but that doesn't mean we need to do them. "Society may come calling. You don't need to answer the door . . . I also joke sometimes you can purposely be bad at them, so nobody will ever ask you again. I learned that from my brother, who used to break dishes on purpose so he never got asked to wash dishes."

COACHING YOURSELF

Beam finds it difficult to speak up and raise questions. "I was raised in a very traditional Asian, patriarchal household, so talking back or speaking my mind wasn't always welcomed with open arms by my family." Now, asking thoughtful questions is a demonstration of cultural fit and leadership at her company. "You need to question what you're doing. You're not going to get a brief, and the brief has all the answers, and you're just going to execute on the brief. You have to investigate. You have to see if it's even the right assumption." Stepping up to own the problem space is key, and while she's always been resourceful, she still finds challenging authority difficult. For instance, "speaking up in a meeting with senior people, especially if there's already a comradery among a set of managers, or there's past history with some people so they'll listen to some people over others. It's been personally challenging for me to feel like I have a seat at the table and feel like I have the privilege to say something while sitting at the table."

To counteract this, she's very self-aware and pushes herself constantly. "I've had to mentally say, *Okay, don't be afraid to ask a question. You're not going to get shut down and you're also not going to get looked upon as stupid. Just ask the question or challenge what you're hearing.*" She's had the experience where she's the only woman. "I think before I would've been intimidated, and now I'm like, I'm here too."

Wendy's done the same thing throughout her career as she worked her way from being an aesthetician to moving to tech to ultimately becoming a security engineer. "I'm pushing myself to ask questions, and I think that's really hard." She recently was speaking on a panel discussing how she got into her role in security. She made a point of telling people to ask. "I am here and they hired me, and I need to be able to be comfortable saying, 'I was good enough to get this job,' but I also need to be confident enough to say, 'Hey, can you explain that to me?'" Even though she was making these large career pivots, it took her a while to get comfortable speaking up in meetings. "I think that was a huge hurdle for me to get past, to be able to admit that I didn't know." She would leave meetings pretending she knew and "then I'd be like, I have no idea what they talked about at all." Thanks to supportive team members, she's become more confident. "I'm like, 'I have a really stupid question: Are these ports that you plug in the same as a port on the internet?' They're like, 'That's not a stupid question,' and they totally laid it out and explained it to me. And it is such a good feeling to have a team that I feel like I could ask." Ultimately this helps her feeling of belonging, in addition to her courage, because she's able to say, "'I have no idea what that means,' because no one's going to judge me and think, *Why is she even working here?*"

AT HOME TOO

When working through her work/life balance with three children, Adrienne found that working from home one or two days a week was crucial. "I did the same thing after baby number two as baby number one: I had one work-from-home day." She asked her manager for an additional work-from-home day, increasing it to two days. Later, when she received a new manager, she presented it as, "This is what I do." The old manager backed her up, and the new manager supported the arrangement. "And then I've just kept it. I've built out my team and I'm like, 'This is what we do.' And everyone's like, 'Great. Work/life balance.'"

Adrienne also found it critical to get help at home. Another mom told her to "write down everything that I don't like to do." Whether it's laundry, bathtime, dinner, "write down what you don't like to do and literally put that into a job description and hire someone to do it for you and then spend the rest of your time doing the things you like to do." She notes that money is the real limit, so this may not be possible in all cases. However, she recommends the exercise and thinking about how you could get help, whether hired or not. "And that was a game changer, because I did that and then we ended up hiring some help in the evenings, and I was like, 'Okay, you're going to do these ten things I hate.'" That's been another reason she can cope. "I can just be. I come home, I walk in, I hang out with my kids, and the dinner's ready. And it was like, okay, all right, we can do this."

Yolanda has also needed to ask for help at home. "My husband works at Facebook, so we're also a double [paycheck] family. We have a nanny come in each morning for four hours to drop the kids off, to prepare dinner, clean the house, do laundry, and grocery shopping." Having help with the kids and house cleaning is "part of our survival pattern."

A quick note: often the complaint about "get help at home" advice is that not everyone can afford this route, so I'm not making the "pay someone" piece a big part of my recommendation. That said, the more we can ask for help, accept it, and realize we're worthy of it (see more about that later in this chapter), the better. It takes a village, and the more we can rely on others for assistance, the more we can achieve both at home and in our careers.

RESOURCES

Ask for More: 10 Questions to Negotiate Anything *by Alexandra Carter*

I didn't expect to like this one because I thought it would only be about formal business negotiation. However, Professor Carter does an excellent job reframing negotiation and its impact in our lives. She leads us through understanding ourselves and our desires before asking others to do so and fill our needs. Also, the metaphors that frame the book, the "mirror" and the "window," are effective and useful devices for how we ask all questions.

Ask for It: How Women Can Use the Power of Negotiation to Get What They Really Want *by Linda Babcock and Sara Laschever*

Linda Babcock kept popping up in the references as I read articles, so I took a closer look at this book as well as others she's cowritten on negotiations and the gender divide. This one is rich with stories and advice, and it also extends beyond what I think of as traditional negotiation, helping me think through what I want out of life.

FIND SUPPORT

It took me a while to realize I needed help, and I feel pretty silly about that, in retrospect. I was a fairly independent kid, which converted to me being a pretty independent adult. I was the type of person who would pack up my own car and move across the country. I would then move myself for the next few years from rental to rental, until I realized that movers existed and I could afford them. The first time I paid for a house-cleaning, also years later, it was so amazing I thought I heard angels sing. Despite this, it took me forever to ask for career help and longer to really listen to the advice. I actually don't think I became a real pro at it until I started the process of writing this book, where it was clear I knew nothing and needed help from people who'd written books or were familiar with the book industry. Until then, I thought I should know it all and really only flirted with getting guidance or help.

But ugh, I was falling into such a classic trap. If "knowing is half the battle," the "not knowing" is the other half. From salary differences to promotion criteria, being in the dark keeps us from being able to achieve what we want, whether that's equal pay or the plum role. Finding support is one way for us to navigate lack of insight or knowledge. That's why groups like Ladies Get Paid have sprung up to provide courses on everything from negotiation to knowing your worth.[41]

As I interviewed women, some sounded totally alone as they navigated their careers. They talked more about their worries and insecurities than the women who had found people to talk to, to share experiences with, and to bounce ideas off of. This makes sense. Self-doubt can creep in when we are making decisions by ourselves, when we have no one to

---

41 Ladies Get Paid. 2020. "Education: Kickass courses to help you get ahead in your career." Accessed June 3, 2020. https://www.ladiesgetpaid.com/webinars.

help us think about whether to take that new job or whether to ask our manager for a raise. In particular, finding a peer group to understand what's typical in salary or compensation analysis can be critical in terms of financial considerations.

Support can come in many forms, whether through coworkers, training programs, or internal email groups. Some are outside of work, such as Lean In circles, women in tech groups on Facebook, or blogs. The phrase "found my people" came up more than once in my interviews, also implicating "belonging" in our set of needs. Whether networking at conferences, finding the right team, or joining a good company that matches our values, women often felt immense relief and satisfaction finding their place among others who understood and supported their path. The support we receive can also extend beyond job topics as women navigate the full spectrum of their lives. Often women are seeking kinship as working women with in-laws, children, and partners, or asking for recommendations across needs for doctors, lawyers, and more.

FINDING EACH OTHER

Amy can't recommend finding support enough, whether through colleagues or organizations. "I like to build a support group around myself and around other people. There's always support out there. You know, there's ten million women in technology, organizations, and summits." She gave the examples of Lesbians Who Tech, the Out and Equal conference, and Women in Technology summits she's attended. She's also leveraged her Wellesley alumnae network and Lean In circles, having launched Lean In within her company as well. "I can't imagine how in the world anyone gets by without reaching out to others, reaching out to your network." She believes life is hard enough without leveraging each other for what's difficult or to learn from what others have experienced. More important, talking to other people can help you see things you might not and help you feel more confidence.

Ashley found a Girls Who Code conference inspiring, the first women in tech event she's attended. She'd hesitated to go as she wasn't sure what to expect, but sitting in the conference hall full of other women was a worthy moment all on its own. "I remember I was looking around, and that moment struck me. This is the first time I've been in a room with this many other female software developers in my entire life." She noted how women were by far the minority in her classes in college, so "I was like, 'Wow, this is really awesome.'"

Aside from the numbers, she also found the content inspiring. For instance, there was a talk about imposter syndrome. "I think one of the things that stuck with me, I don't remember it word for word for word, but the general gist was that everyone has their own experiences and point of view. That point of view is important, and we were bringing something to the table that no one else can bring because they don't have that same experience as you do, and you're going to see things that other people will miss." Ashley mentioned examples like offensive commercials that might have benefited from a more diverse team. She keeps this in mind. "I try to remember that I have a unique point of view."

Cathy is happy that there are organizations like Girls Who Code and organizations for minority women, "affinity groups to find others and to get that sort of support and official, formal mentorship programs." She didn't have those as she was embarking on her career, and at the time she didn't think about it. "I was like, this is the way it is. I'm going to deal with it." While she feels she dealt with it fine and even thrived, she does think it's important "being a woman or a woman of color in any corporate setting or any job, to find those allies." Allies may come in any shape or color; what's important is to "make sure you're not isolated and having people as your sounding boards to help you in a career or in personal stuff." She gives the example of her coworking space, the Wing, which encourages networking. A woman recently reached out via the Wing's internal messaging app and asked Cathy to be her mentor. While Cathy was unsure at first what she could offer, this woman was earlier in her career and was having trouble speaking up in work presentations. "Some white males [at her job] were really big talkers and good at talking and good at using big fancy words that got their clients really hyped up. She said, 'They're smooth talkers, they're salesmen, and I don't feel like I have that skill. People don't take me seriously like they do them.'" Cathy encouraged her to leverage "other skills and abilities that you can use that they can't," noting that the mentee was detail oriented and honest. Cathy suggested leveraging those elements of her nature as a way to build trust with her clients.

Wendy had a negative experience at a security conference once when a man wouldn't give her a T-shirt because he didn't believe she was a security engineer. She reminds us, "There's a lot of people that don't believe in you, that maybe will discourage you." Her advice? "Don't let those people be there. Your tribe is somewhere. Find those people that lift you up and encourage you, whether it's family, other coworkers; find people with meetups. But don't let that guy at the conference that tells you you're not good enough define what you do next."

PICK YOUR BOSS

Kristen says, "Really pick your boss." When she thinks of the support we need, that's her primary thought. "As you're interviewing, especially early in your career, you're so excited and motivated to get the job. But in some situations you really want to make sure that you're interviewing your boss as much as your boss is interviewing you, because who you work for totally dictates your experience at a company, or at least that's been my experience." As a mom of four, Kristen has a specific list of questions she asks during job interviews: "I always ask if their husband or wife work. I have found that working for someone whose spouse works is very useful in their understanding of how my world works." Since she'll need flexibility, she wants to see her boss also naturally understand that versus having to teach them later. She also asks about "their decision-making, like how they think about things, how they solve problems, failures they've had. I try to understand how self-aware they are, what their visions are." Seeing what makes potential supervisors tick helps her understand how visionary they are and how they'll approach their daily work environment.

Elizabeth's advice after having "a very positive experience overall in all the organizations" she's worked with is that "success is a combination of your skills and your setting, and you have to make sure at a certain point you are interviewing the company and [determining] whether or not the setting is right for you." She is clear that the responsibility is ours, and we can't outsource responsibility to a company. "You have to do your homework and really think about, is this a place that will allow my skills to blossom because of this particular setting?"

When asked how she navigated this and what her advice would be for others trying to pick their opportunity, she too said, "The key is really understanding who your manager will be." Similar to Kristen, she wants to understand the boss and their process. "I like to really understand in some ways how the day-to-day works. How do you keep an eye on things? Do you have one-on-one [meetings]?" These questions help her understand "how the day-to-day operating rhythm works, because that really determines your success."

FIND THE RIGHT CULTURE

Yolanda Chung also recommends looking beyond the manager. She hasn't enjoyed all her experiences working in tech, and she found the start-up culture in particular not a great fit. "The leadership and culture there definitely was not as strong, and so that was not a very fun experience at all. It was very much your stereotypical, very not friendly to women kind of experience." Her advice having dealt with that is to explore who you are working with, not just the manager. For example, seek out intel from fellow employees via resources like LinkedIn or Glassdoor ratings.

Camie concurs. "Look, the culture matters. Go to an inclusive environment. That's the most important thing for being happy. A lot of people choose based on the project or [other factors]. I have the benefit that I've got thirty years' experience, I've done all kinds of projects, I've done start-ups and big companies. I can't even count how many projects, how many products I brought to market. . . . At this point, I know that it's the people that make the difference in whether I'm going to be happy or not."

RESOURCES

TableTalk: Hearing the Silent Fear and Bridging the Gap *by Shari Moss and Meghan Fitzpatrick*

This book was a pleasant surprise when I stumbled upon it while searching for resources. While targeted toward millennials, it has solid cross-generational advice for how we build our own board of directors and its importance at various times in our lives. It made me think about who my board members are!

A Mindful Career: How to Choose a Career, Find a Job, and Manage Your Success in the 21st Century *by Carole Ann Wentworth and Eric C. Wentworth*

Wow! This book is basically an encyclopedia of everything to know about the current job market, how to target the right career, and how to land a job. In particular, I appreciated that the book included research to do during interviews about your employer and its culture, including a chapter named "What You Wish You'd Done Before You Said 'Yes' to the Job."

Taking the Work Out of Networking: An Introvert's Guide to Making Connections That Count *by Karen Wickre*

Karen Wickre was a "weak tie" of mine, her term for the people we know slightly but we still contact for guidance. Despite not having seen each other or spoken in years, I reached out to Karen via LinkedIn when I was preparing to write this book to see if she was open to discussing her book writing experience. That led to me also interviewing her for this book, given she is a former tech employee and now a consultant. If you hate networking, this book's for you. Karen systematically walks through ways we can connect with others, especially when it's not our comfort zone.

Please also refer to the Appendix for "Mentorship 101," also available on alanakaren.com.

OWN YOUR AWESOME

When I was writing this book, multiple people asked me what I wanted readers to take away. While there are many stories, ideas, and tools covered here, I'm writing this book mainly so you'll feel like you belong in tech and that you bring an awesomeness that only you have. I deeply want you to embrace that feeling early and often in your career.

Why? While I'm not much of a conspiracist, I do believe that women are generally taught throughout our youth to think small and doubt ourselves. We're fed early the idea that the littlest things could be turned into insults against us—what we wear, how our hair looks, whether our schoolwork is tidy, whether we get straight As while being nice, whether we act too loud or too bossy. We are pushed back into our boxes time and time again, and that's convenient for society. Someone needs to tend those boxes; having us be the responsible caretakers has been a practical need for generations. And while that's starting to change, it's not going to evolve overnight. Likewise the intersectionalities we may experience as other kinds of minorities have their own histories and implications.

The downside of this? No less than economic opportunity for our governments, our families, and ourselves. Quick facts from the UN Women website[42]:

Women's economic empowerment boosts productivity [and] increases economic diversification and income equality in addition to other positive development outcomes. For example, increasing the female employment rates in OECD countries to match that of Sweden could boost GDP by over USD $6 trillion, recognizing, however, that growth does not automatically lead to a reduction in gender-based inequality. Conversely, it is estimated that gender gaps cost the economy some 15 percent of GDP.

Putting aside the larger societal impact, how you feel is more important to me. Are you plagued by insecurity? Do you feel like work is taking

---

42 UN Women. 2018. "Facts and Figures: Economic Empowerment." Last updated: July 2018. https://www.unwomen.org/en/what-we-do/economic-empowerment/facts-and-figures.

advantage of you? Do you want so much more out of life? Do you like yourself? As I speak with women, it's common that we wish our younger selves knew to question themselves less, to embrace our talents more, to not let anyone push us into a corner, to ask for what we want and take it. I can only imagine what the world would be like if we knew this earlier—if at twenty, we realized we could own the room instead of at thirty. If at twenty-five, we thought we could lead a team versus thirty-five.

The stories below cover the gamut of where women are on this journey, what they've learned about themselves and what they still need to learn. I wish for all of them, and for all of you, I could wave a magic wand and box up the feeling of being awesome and give it to you. That's a tall order, though, so you're going to have to read the stories below and then think about times when you've felt a similar way, where you've come to the same realization, or where you've wanted more. And then, starting tomorrow, try to remember each day that you are awesome, and you don't need anyone else to say it for you to believe it.

MY OWN STRENGTHS

Ashley has found confidence through having other hobbies and embracing her unique talents. "I have a balanced life and have other hobbies. I think it's good, and that's something that I actually bring to the table and gives me a different point of view." Remembering that and garnering some experience has helped her. She tells an anecdote of being praised for being able to work with difficult people ("not in terms of being mean to me, just kind of grumpy"). She's found strength in getting to know those people and figuring out how to approach them. "I know I'm not the best coder in the entire world," she says, and likes that she balances it with people skills, problem-solving, and being independent. "I figured out over time I have my own strengths."

Annie finds confidence in knowing the language of where she's working, whether that's technical or team-specific lingo. As she takes on additional or new roles, she dives in and figures out how to contribute more. "What classes should I take on this part of their tech stack, or what can I do to learn more here?" At the same time, this isn't being technical for tech's sake. Annie has found it critical to know her own strengths and limits. Part of why she wants to know the language is to be an effective translator, one of her key strengths. "I'm like a diplomat translator; there's so many different ways of naming it. And then understanding, what empowers that? What gives you more of a voice?" Since she's on engineering and product teams, investing in skills that help her be a strong translator not only helps her teams be successful, but also enhances her own reputation by elevating her voice and value to leadership.

Sara remembers hearing a speech I gave that borrowed straight from Shonda Rhimes's book, *The Year of Saying Yes*. "I remember . . . one of the things you were going to start to work on was, when someone compliments you, to say thank you and nothing else. And that's something that stuck with me." She often likes to step back and reflect, but she sees that she'll focus less on appreciating herself and thinking about what her strengths are. "I know for myself, I spend all the time thinking, *How could I have done this better? Where do I want to grow? Where do I want to improve? Where do I want to advance?* And I spent very little time thinking, *How did I kill it today? How did I do a good job? What about me makes me successful?*"

BE TRUE TO YOURSELF

Belvia shared a story about how she reported to a VP earlier in her tech career. She kept asking for feedback and was clear with him that she preferred direct, honest feedback, "even if it's going to hurt my feelings, it's okay because it's something I need to fix." The VP would say she was doing great but then he would complain about her work to others. When she confronted him, he brushed it off like, "I was just having a conversation with that person." Over time it made her second-guess herself. "Then you're making more mistakes, and then there's this angst. And I remember being so stressed out, I actually threw my back out." Ultimately she ended up switching roles to be true to herself and what she needed. "I had never been that stressed out over something before, and all because [he wasn't] honest."

Per Belvia, "I would say we have to learn to be true to ourselves. Don't let other people put their angst and how they compartmentalize people on you." She wants us to remain on our path regardless of what others say. "You're always going to have the naysayers, and you have to learn to tune them out." We spoke about why it took her a while to leave that position, and she credits age and maturity for how she'd react now. "I probably would've left at six months, knowing what I know now." She thinks some of this comes from being a woman and having a nurturing side. "I think that need to please that I have and that need to want to make things right is where I was really trying to strive, and I should've cut it loose and been like, *this will never work, so why am I going to waste all this energy*, but now I know that."

Sheri echoes a similar feeling. When I asked how she's navigating proving her worth, she answers, "In all honesty, I think it's getting to the point where you don't actually care." She gives examples like "you don't mind saying, 'Hey, don't talk over me' [or] 'Wait, you just said exactly what I said.'" It's about reaching that point of knowing your worth and not caring that you have to assert yourself, "practicing and then realizing that you actually do have a place at the table."

OUR OWN ENEMY

I noticed a pattern as I spoke with some women, one that I'm familiar with in my own experience. Occasionally we can limit our own horizons, hold ourselves back. When asked about her career challenges, Belvia talks about herself as her primary blocker. "There's always that self-doubt, and when things come up, changing jobs and having all these opportunities, I've always felt that I'm not supposed to have it." She links it to her upbringing. "I didn't have the worst childhood, but my mother . . . like, I am one generation away from Jim Crow, and my mother grew up in that Deep South." While she had a home economics teacher pushing her toward fashion, "then I go home and my mother always told me what I couldn't do because I was a black, or they don't let black people do that, or why would you do that? Because you're just going to get hurt. So why put yourself out there for those kinds of things?" With positive people around her and having learned to quiet her negative inner voice, she's found ways to move forward. She credits friends and a broad set of mentors for helping to gut check these thought patterns.

Similarly, Ching has been working to block out her own inner critic and not fuel her own insecurities. While people in tech might have preconceived notions about her or her role, she calls out the ways she holds herself back. "People have said, 'You are a woman in tech.' I'll be like, 'No, no, no, no, no, no. I don't code. I am not an engineer.'" Over time, though, she's become more comfortable wearing those shoes. "It's taken some time to realize, no, I am a woman and I am working in tech, and I am contributing in a significant way." Her negative thoughts can still outweigh her positive ones. "It's that inner voice that I've had to learn how to strengthen over time, because I also have learned over time that I have a supershouty inner critic." That critic will hold her back. "I will tamp myself down perhaps even before somebody else in the room will do it." Now she has to catch herself before her naysayer takes over. "I literally have to shout over my own voice just to be able to maybe say something in a meeting because I am second-guessing myself."

INSECURITY

One of the real threats to our personal and career growth is the insecurity inside each of us. Laura talks about how her own insecurity has come and gone over the years, especially "when things are constantly changing and you're moving a million miles a minute, it can feel very chaotic and intimidating to have an opposing opinion or big idea that others might not agree with." She feels like this has been related to both her gender and age. In her role, "I've been the youngest on the leadership team and, for all but 1.5 years, the only female. Some days I'm feeling incredibly confident and able to ensure my voice is heard and that I'm also educating everyone on a different viewpoint, but other days I feel I need to prove my value and justify my voice in the room." She hasn't been able to pinpoint what drives her feelings. Is it something happening in the room? The context of the day? A shift inside herself? "I'm a very emotionally driven and self-aware individual and often put myself out there far more than others, in a way that makes me quite vulnerable." One culprit may be if she puts herself on the line, but others don't return the same energy. "If others around me aren't doing the same, which I would argue is key to building a strong leadership team, it can feel strange and like you've just opened up in a very raw way and [been] made to feel like it was inappropriate or unnecessary." Laura finds that working with an executive coach has helped her via mindfulness and meditation practices, but she's still figuring out how to find time amid other work/life balance pressures.

YOU ARE WORTHY

Maybe you still don't believe in your awesomeness in your career, and I get that. Maybe start first at home? Beyond being worthy of career opportunity and a voice in the room, I also want women to feel that they deserve boundaries, rest, and a healthy home life. The need to sacrifice ourselves at all costs is not sustainable.

Georgia is thinking about how to set up her life for a family and work right now as she has her first child. "I think I have to know what kind of space that I want. What would make me happy? Ask for that." She's referring to the kinds of boundaries or jobs she'd want to have, with fewer on-call requirements and 24/7 emergencies. Once she's done the work to figure out what she wants and ask for it, then she needs to stick to it. "If I do get granted those things, [I need to] hold the people accountable to actually let me do them." She knows that she's historically chosen to work after hours, something her husband calls her on. "That is a choice that I am making, to respond, and that feeds this cycle and incentivizes people to continue to keep doing that." She now views it as her responsibility to end that cycle.

Annie spoke of a similar idea: Where are we our own worst enemy in finding balance? "I think that tech is something where you have to put your own boundaries in, and it's very easy to say yes. I think also, as a woman in tech, it's very easy to say, 'Yes, I will do this. I will take care of that. I will clean that up.'" She admits that saying no and setting boundaries is hard, especially when there's a major initiative like going public or a product launch. "No one will tell you to stop but yourself. And tech will not tell you to take a break. There's unlimited vacation, which means never." In addition, she feels like there is more to prove as a nontechnical woman. "I do feel like as a woman in tech, it can be harder because there are fewer of you, and sometimes you feel like you might have more to prove. I think also, coming from a nontechnical background, I feel like I have to prove myself in other ways."

Ultimately she took time off after feeling burned out at a previous company. "It was very good to reevaluate—What are the things I value? What are the things that I want to build?—and come back really energized." Now she thinks through both when she turns on the computer and how she feels. "You have to just shut the laptop off and also understand what I did is good enough. I have this expertise, I have this experience, I have this knowledge. And that's good enough for today."

Laurie came to the same conclusion: setting her own boundaries has been key to her finding better balance. "What I've learned is people get stressed out and they expect their manager or their team or the leader to set boundaries for them, and really you're the only person that knows your own needs. So you have to set those for yourself." She is careful now to not respond to email late at night. "Sometimes I'll still look at my email and I'll make a mental note. I'll [flag the email], but I will not reply back, because my boundaries are my hours." She now knows that taking a break is key to her performance. "This is my time. And this is what I need from a mental and physical decompression state for me to be the best employee that I can be."

After being in some roles that were demanding in terms of long work hours and commutes, Natalia now seeks roles where she can both have a personal life and love her work. In particular, she feels like being able to detach from work is critical to her performance. "I'm getting better outcomes than ever before. I can connect dots more when I add more balance; it's not always about throwing yourself at a problem. Sometimes it's about trying something, stepping away to relax, and then coming back to the problem; then you're more effective." This was different earlier in her career. "I didn't know how to say no to my manager when new work would be added. I didn't want to, and I was very grateful for the opportunities." She's also had to dispel her imposter syndrome, which was driving her to work harder to prove herself. She leveraged an internal support group which helped her see how her manager at the time was feeding her feelings, eventually leading her to get support from mentors, her skip-level manager, and her HR business partner to resolve a challenging situation. "I think if women believed in themselves more or realized they didn't con [their company], that feeling imposter syndrome is just part of the process, they might get to a more productive and effective place quicker."

She now thinks about the business impact or context before taking on additional work. "I'm very good now I would say [at] putting up boundaries, like prioritizing things, saying no or saying 'not now' or saying 'I will add this to this tracker and we can look at it next quarter.'" She is aware this approach might not work for every team or environment, however. "A manager in my previous role told me that if I wasn't willing to do the seventy- to a hundred-hour work weeks, it wasn't the right role." While she acknowledges some people will want that, it's no longer worth it to her. "It took me a while to realize that it's not that [the company] and my managers were doing me a favor. This is a partnership, you know?"

RESOURCES

The Self-Confidence Workbook: A Guide to Overcoming Self-Doubt and Improving Self-Esteem *by Barbara Markway, PhD*

If you've done a lot of therapy already, you'll possibly find this one repetitive. Otherwise I recommend it as a step-by-step guide to identifying negative patterns in how we perceive ourselves. It includes action items, key takeaways, and confidence-booster exercises.

Daring Greatly: How the Courage to Be Vulnerable Transforms the Way We Live, Love, Parent, and Lead *by Brené Brown*

Brené Brown has achieved near-guru status, which is always double-edged; some folks may not like her approach and all that talk about shame and vulnerability. That said, if you want to dig in and understand the triggers behind our emotional reactions and deep, dark self-beliefs, I'd start with some Brené. Take a look at her TED Talk videos online if you want to try before you buy.

Year of Yes: How to Dance It Out, Stand in the Sun and Be Your Own Person *by Shonda Rhimes*

I was drawn to this book to learn more about Shonda Rhimes, a successful creator and producer of television shows and a supporter of diverse talent amid typically unsupportive environments. Her personal journey turned out to be extremely compelling as she figured out how to love and accept herself. I read this years ago, and some of her insights (just say thank you!) have stuck with me to this day.

I was nervous writing this chapter. How could I possibly contribute more to the already crowded career advice space? Should I even try to do this? Then I took a deep breath, I asked for advice, and I found support in the research and wisdom of others. I chose to believe my advice was valuable, and I remembered that my clear-cut style of explaining tough subjects helps people. This is how I beat insecurity. Bit by bit each day.

But it's hard to do alone. The next chapter, "Our Champions," shows why mentors, sponsors, and other helpers are so critical in our career journeys.

9 | Our Champions

When I think back, despite countless helpful people in my career, there were two women who took the biggest chance on me. They were my sponsors.

Much is made of sponsors in today's work environment. They aren't simply mentors who give you advice or coaching. Sponsors lift you up and find you vital opportunities. Sometimes they believe in you more than you believe in yourself. Sometimes they have the power to make magic happen, to accelerate your career in ways you never would have been able to do alone. They can be men or women, and I've had both act as my sponsors, go to battle for me, and help me get promoted. But it was two women who saw something in me and took the biggest leaps.

What if they hadn't been there? Maybe I would have still survived on skill alone; maybe I would have kept surprising people in meetings, watching their eyebrows lift and respect dawn over and over and over again. I believe in myself; I didn't need them, but they helped me exceed my expectations. I didn't have to wait and convince; they raised me up on promise alone. They reassured me when I was second-guessing myself. They leveled the playing field.

I was in my second year of college and nineteen years old when I walked into the International Center at the University of Virginia and applied for an internship that usually only accepted third- or fourth-years (UVA speak for juniors or seniors). Housed in a beautiful old Virginia brick house, the International Center helped facilitate and welcome foreign students. They also hosted cultural events like Bengali Night to encourage students to share in each other's foods and traditions. The goal was to give "foreign" students a home away from home while also exposing "domestic" students to a world outside of their own. As something of an outsider even in my own country, I liked helping others. The internship was also for school credit, and I had already figured out that I preferred learning by doing.

Lorna Sundberg had long ran the International Center and had met many students in her time. A tall woman with clear eyes, she was no fool. She kicked our discussion off by saying she was very doubtful that a second-year student could handle this internship. As a whole, we lacked skills and maturity when dealing with sensitive situations like newcomers to our country. I liked challenges, especially when people doubted me, and my dad hadn't called me "thirteen going on thirty years old" for nothing. By the end of our conversation, she leaned back. I could tell I'd impressed her; she later confirmed she'd take a chance on me.

I worked at the International Center for the next three years. I ran countless events, showed new students around. Lorna even humored me and let me update their website. I worked side by side with her and other employees, and the Center became a second home to me as it had for students from all over the world. That experience led me to other web-based internships and ultimately my webmaster job at the university after I graduated.

Lorna was my first sponsor. Today the International Center is named after her, for her wonderful work fostering the world's students, including me.

Sheryl Sandberg was my second sponsor. She's now famous, but at the time she was simply my boss. I was twenty-four years old, and I had recently joined Google. Shortly after I joined, Sheryl took over a small team managing ad customers and inherited me in the process.

I'd met Sheryl previously in our little gym. I was running, and she was biking. I was training for a marathon, and Sheryl was putting herself through her paces, as I imagined, and later learned, she did every day. Sheryl is a tough cookie; I felt her high standards for herself and others at that first meeting as she grilled me on my background and what I was working on while we worked out.

But somehow, I won her over. I was hardworking and smart, but her regard seemed like a magical and undeserved gift at the time. She visibly took me under her wing, inviting me to sit in on management meetings and taking me to senior executive meetings with Google's CEO and co-founders. She opened up my career for me through her belief, frankly dragging me up the ladder with her, given how fast I was attempting to grow and onboard skills. In return she had my utmost commitment, and I worked euphorically and unendingly during those early years of my tech career.

In March 2008, Sheryl left Google. When she first told me she was leaving, I took it stoically and maturely. But when I was saying good-bye a couple weeks later, I burst into tears. We kept in touch, but even then, I knew it was the end of an era. In that moment, I was losing a major sponsor—or in more human terms, a major believer in me—and I would not fully understand until later what that really meant.

What began at that moment is a period most easily called "when everyone who thought I was awesome left." This is an exaggeration, of course, but also pretty true. For the most part, the people who had followed and believed in Sheryl's leadership or who had been fostered by her presence left Google in the years that followed. And those were the people, many of whom were friends or close colleagues, who had witnessed all my hard work and were my biggest fans.

Up until that point, my career had been on an admittedly extreme trajectory. I started at Google in 2001 in an entry-level customer service role and was made director by 2007. I worked very, very hard (and frankly I barely remember my personal life as a result) at a time when business was growing very quickly. And I also had great sponsors watching out for me who rewarded and pushed me even further. It was the best-possible career scenario, and I knew it.

That changed overnight, and the most noticeable part was that I no longer felt awesome because I wasn't recognized as awesome anymore. I probably could have followed my sponsors to other companies and had an easier go of it, but I didn't want to. I wanted to find out where I would go on my own and what kind of leader I could become. And I fundamentally believed in Google and its mission—no company had ever appealed to me more.

So I stayed. It was a deliberate but painful journey, and I've been frustrated and downright angry at times. But it's been great in so many ways. I looked up from the day-to-day and really focused on what type of leader and manager I wanted to be. I fully explored my role to its very limits, fortifying my strengths and plumbing my weaknesses. I had three children, which brought its own set of opportunities and challenges. And I remembered all on my own that I was awesome and didn't need others to tell me so. But what if I hadn't? Throughout my story are major leaps, bets, and turns of luck. What if I hadn't had two great sponsors? What if I hadn't fought to learn technical skills without help or classes? What if I had taken disinterest as a sign I wasn't worthy? And what if those women hadn't been there at that time? My story would have been very different.

My sponsors are not a story or a movement to me. They are real people who reached out and gave me opportunity when they did not have to. Did my sponsors have to be women? No, plenty of sponsors are men. Thank you all; please keep going. Despite that, I feel strongly that we need many types of people in leadership roles in order to diversify who our sponsors are—we simply see different things in each other, and the more variety we have in perspectives, the more that different kinds of people will achieve the help only sponsors can bring.

I was curious whether other women had similar experiences, but interestingly I didn't start off asking about sponsors or mentors. Given that there is a lot of talk about the importance of sponsors and mentors nowadays, I didn't want the usual sound bites; I was more interested in where the stories would naturally emerge. For instance, I asked people about a key career challenge, and then I would ask a follow-up question about how they navigated it. Often stories of their career advocates and guides emerged there. As I listened, many women seemed alone in navigating the ups, downs, and decisions of their careers. I started to pointedly ask those women whether they'd had mentors or sponsors. Below is a collection of comments that emerged across what women needed and how they received help.

THE SUPPORT WE NEED

Olivia and I discussed the belief and help that can come with having someone who brings you along with them up the ladder, even though she's still figuring out how to achieve it for herself. "How did you get to where you are? Did you come with someone, alongside them? Did they pull you up as you moved? Did you do it by yourself?" In her case, she never felt that support. "There never really was someone that was like, 'We believe in you.' It always was whatever I got, it's sheer will, determination and resilience, anger—all of those." She wonders how you get the person who "trusts in you that lifts you up" and how you avoid the "someone that you pissed off, you didn't mean to, and they will boycott you every step of the way." She's seen this play out with friends' careers. "They have been told, 'You are not voted off the island, but you're definitely voted off this boat.'" How do we know when to move on and when to fight if we aren't using advisors as sounding boards? "Because you've been told all your life—maybe not all but some of us—I've been told, 'You're a fighter, you don't give up. Fighters don't give up.' At which point is it?"

Another woman shared how critical allies would have been as she was negotiating salary. At the time she started at her company, she took less stock for a slightly higher salary due to cost-of-living needs. On her first-year anniversary, she "wrote a thank-you note to my boss on monogrammed stationery, telling him how much I liked my job and how excited I was to be there. I didn't know that you were supposed to ask for more stock. Everybody else did. Nobody told me. I had no idea." Given how well the stock later performed, this was a big deal. Having an advisor or peers to share that kind of information with you can be crucial.

Today she leverages a crew of people she's close with to bounce ideas off of, but she does find it hard to find women in similar roles as people's careers and home lives have diverged. She's recently been leveraging an external mailing list group for support and resources as well. "It's a group of diverse women. They're all working. Some of them have kids, some of them don't. So sometimes the kid stuff bleeds in and out, but it's been a great outlet to say, 'I got mansplained again today.'"

Jennifer T. also wants us to focus more on the bottom line and consider sharing sensitive information to benefit ourselves. "Gloria Steinem once said that the most revolutionary thing that women can do is talk about money. And I think it's really, really important, especially when it comes to our compensation." With a notable gap in relative earnings between men and women, Jennifer thinks there's power in sharing what we make with each other. "We talk with our closest friends about the hardest things in our life, and I've never sat down and talked to my best friend about what I'm doing with my money and how I can be better investing or better supporting myself and my family." While she knows women want to be able to focus on doing a good job, she knows that's not enough to help us really thrive. "I want women to talk more and have hard conversations about access to all of it. Whatever that is, whether it's that job, or that entry into tech, or [understanding] the jargon, or the skill set that you need, the mentorship, the funding, how to better know what your base salary should be, and how to have your money work for itself. All the things . . . so that you know what's possible."

VISIBILITY

Leanna hasn't had anyone she feels showed her an explicit path, "but I've definitely worked for both men and women that were strong supporters of my work and of my decisions." She's felt valued by the organizations she has worked for, and she's seen them "raise me up to do what I wanted."

Adrienne felt like the help she needed was about seeing opportunity that she couldn't if she wasn't in the room. "The feeling was all the men are running this; they get in a room and do their leadership off-site. I'm on the leadership team, but everyone above me on that exec team is all men. And so it was just a little bit of the boys club feeling." While she didn't feel like they were doing anything to prevent a specific opportunity from coming her way, she likened it to the frog in boiling water, where she may not be able to see what's happening. "Is this really happening? I don't know. I can't see."

Belvia thinks of her great bosses as the ones who have listened to and amplified her voice. "I've had probably three horrible bosses in my entire career span, but they don't outshine the [good] ones that I've had because they would always let me have a voice, and they've never made me feel less than or stupid." They've been the ones to credit her ideas and try them out. "So I've always felt like I've had a voice at the table. I mean, I'm not high management or anything like that, but for my level and where I've been, I've always felt that I've had a voice and that it's been listened [to], that I've been heard." Her current boss is one of those gems, and she credits that along with the access he gives her to learning. "I think the reason why I've stayed so long is because I've learned so much from him. He's always letting me have a voice and learn."

ROLE MODELS

Ashley wants people to see more diverse examples of success. "I think the misconception about tech is what people see in the media, where the software developer is insanely genius, does it all day and night, and is coding all the time and working on all these side projects." When people think she's that type, she explains that coding is difficult for her, and it's not like "I can speak to computers now." She wants to serve as an example of someone who persevered, especially to those who are struggling. "If you're doing worse than your peers, or you feel like you're not understanding anything, or if you're getting bad grades, I don't think that needs to mean, 'I have to quit because I'm bad at it.'"

Part of why she thinks about this is that she didn't have a female developer to talk to when she was studying. "I had dude friends, of course, and it was good that I had a group of friends because without them, I don't know [that] I would have gotten through my major," but she sees now how people relate when she tells her story of how it didn't come easily for her.

She also serves as an example of someone who has hobbies outside of tech, for example fashion and beauty. "The tech traditionally [seen] in the media is very nerdy. It's not glamorous. And I'm like, tech life is awesome. You get food and snacks, lots of vacation, so many perks. It's pretty glam, but in people's heads you're stuck away in a basement and super nerdy, and you don't have friends." She's been happy to serve as an example for others. "Other girls look up to you and they see you, and they see you thriving and doing well as a developer and also having other interests. I think that part is also important; like, 'Oh, she isn't what I pictured a developer or a female developer to be.'"

She's also made an impression on men. "Sometimes I get messages from guys like, 'You changed my view a little bit of what a software developer could be.'" She thinks it's comforting to see what she calls the in-betweens. "I think it is important—for minorities, women, whoever—to see those in-betweens. And it doesn't have to be some triumphant story. . . . But I think just being aware that there's other people who share a similar story and it's not one thing or the other . . . I would've loved to see more of that when I was in college. It would've been very comforting."

Bethanie has never reported to a working mom during the thirteen years she has been a working mom. "I see a lot of women, maybe ten to twenty years older, that are single, don't have kids, and this is just all they've known. So I just don't feel like I have had as many tangible role models that are my peers." She's looking ahead and wants to see what her career could look like in the future "to say, 'What does this next stage of my life look like?'" She knows they're out there and wants to find them. This is connected to why she participated in this book, in fact. "I think that's why this project and this book is really important, so thank you."

WHEN TIMES ARE TOUGH

Laurie leveraged support when she was facing adversity, and her supporters helped her get through a hard time. She described how she did a presentation for an executive-level meeting during a reorg (meaning she didn't have a manager at the time). A male leader two levels higher than her pinged her afterward and asked to talk. She explained how this was during a time when many managers were first trying the ideas presented in Kim Scott's book, *Radical Candor*. "Everybody was trying it out, and I say trying it out because people thought it was this trendy thing, but you really have to have enough emotional intelligence to understand how to use it for it to come across correctly." The manager started "reaming" her when they met, labeling it as radical candor feedback but saying things like, "You just ruined your career and you've ruined your brand," and, "You pushed the project back." Laurie felt unworthy and demeaned, thinking, "I just ruined everything. Why should I continue working here?"

Despite that, she regrouped and revisited the *Radical Candor* information. "If he thinks he can give me radical candor feedback, I can give him radical candor feedback on his feedback. I went back to the Kim Scott quadrant [to see] what radical candor actually entails, what it means, and he totally missed the mark for a number of reasons." She describes how if you aren't meeting a couple of criteria, then it's "obnoxious aggression." When she gave him this feedback, "he actually rebutted and was like, 'No, no, no, no, no, I didn't mean that you ruined your career.'" She knows what she experienced, however.

During this time, Laurie turned to an internal support group of similarly leveled women who had attended a leadership growth training together. "People not understanding how to use radical candor in the right context almost caused me to leave. I posted it on [the group alias] and so many women anonymously replied back to me specifically and were like, 'Yeah, I've had this issue' or 'Want to meet for coffee?'" Laurie was hearing this from other women, some of whom hadn't gone back to give feedback and continued to "carry that around." She also received support from a female director who had managed her boss in the past. She said Laurie had handled the situation correctly. The director was aware Laurie's boss was going through personal issues but said that it wasn't okay what he did regardless. She also said Laurie could go to HR if she wanted to and tell them. It was huge "having the comfort of knowing that I didn't do anything wrong per se, it was people just taking out their issues and personal issues and insecurity." With their support, Laurie was able to feel like she did the right thing and remain strong. She continues to have a cohort of women she regularly leans on as well as her last director, who is "amazing, and I still reach out to him for advice."

Natalia likewise has leveraged her community when work was particularly challenging, but she didn't make enough time for this during the first two years at a tech company when she was relatively busy in her role. At one point her role's scope was reduced, "and then one of the first things I did was speak to some people from my business school alumni community . . . and it was definitely getting another set of eyes or another perspective that made me realize where I was having imposter syndrome." Via that community she "crowdsourced some really good wisdom" and realized that she didn't need to be so overworked all the time.

Amy spoke about "having a board of directors [for yourself] that you bounce ideas around and who have your best interests at heart. Oftentimes your mentors and the people in your life think more highly of you than you do yourself. I think we're all really hard on ourselves, and to have people holding up a mirror to you and saying, 'You're the amazing person, this is what I see,' brings you back. We're all so hard on ourselves." Helping others is also never far from her mind as she builds her own network. "There are people willing to help and support one another, and if you don't have that, build it up for yourself. Next step, find it for other people."

RAISE YOUR HAND

Ching has learned from mentors and sponsors the "ask" skill from the previous chapter. "You have to ask for these things, right? There is a lot about raising your hand and standing up for yourself and saying, 'I need help.' And I used to be somebody who had problems doing something like that." She found, though, that with isolation, "you just can't get done what you've got to get done, and you won't . . . you need mirrors to reflect back to you what's going well, what isn't going well." There were times where she would ask leaders for help with her career, and other times where they would step in proactively. "They probably saw that I needed help and I got a little boost, but I would say in most cases I raise my hand: 'I want this project' or 'I'm interested in that' or 'I'm probably going to totally screw this up, but I'd like a chance at this.'" Asking for opportunities ended up being a bridge to mentorship and sponsorship. "I think what I have found in asking for those opportunities, I've built relationships with super senior leaders who can make a difference." She has seen the results of that work in her last promotion. "I looked at the people who wrote [positive comments] for me and I thought, *This is insane*, like, this many senior-tenure [colleagues] going to bat for me, and that made all the difference."

Melanie remarks on how women can be "overmentored but undersponsored, which does not let them continue to grow in the company org chart." By this she means people are always willing to coach or give advice, but women need more than that; they need people who will see their value and sponsor and argue for opportunities for them. As she's been working through challenges at her company and in her role, she's been setting her sights beyond just advice toward meaningful relationships with her supporters. Melanie explained how she's "worked so much more on my relationships and being influential." As a result, she's seeing development. "I have seen so much traction because I think I have found new sponsors [by] influencing the company, and I think it's made a difference for me."

Paola also had to overcome her fears of asking for help and networking. "I think that a lot of times networking is cringy, and for some people it's uncomfortable. I get it. You feel like you're imposing on people." She's gotten more comfortable by thinking of these as "brief connections" to keep in contact with. "We'll touch base here and there, but having those networks with people has been immensely helpful because you never know how you can support them or how they can support you." She'll reach out for a specific question to help her overcome her reservations about taking their time. "Did you have a situation like that? How did you handle this?" This has also helped her confront her imposter syndrome, seeing that not everyone has the perfect answer or that she's not handling things the wrong way.

THE PUSH

Elizabeth talked about the big leap she made to her role in eBay, which might not have happened without a push from a person close to her. She'd had a career in large-volume processing in financial services, but the role at eBay was a finance VP role. While her expertise would end up being a very good fit, at the time she didn't necessarily think she was the right hire. "In retrospect, not that I would actually tell them, I would not even say I was the perfect person for that role. I would have recommended one or two other people in my own company who would probably be more qualified from an operational standpoint. But I wasn't going to let them know. And certainly relative to a finance person, I was far more qualified."

When I told her that I was glad she didn't exclude herself as people sometimes do when they don't meet every single qualification, she said, "Oh, well you'd love this," and told me the story of how she ended up applying for the role. She wasn't happy at her previous company, and her husband was scanning job listings online. "He was the one who actually encouraged me to apply for it because I looked at that [job posting and] said I knew nothing about billing or any of this type of work. And he said, 'No, it's exactly the same except you're doing it for the payment side.'" He connected her with someone who worked at eBay and that led to the role. "So actually, from a traditional standpoint, I did end up limiting myself, but it was him that encouraged me not to."

THE CRITICAL MOMENT

Stephanie and I met at a networking event, and she wrote to me afterward on LinkedIn when I posted a question about when another person served as an example for others in their career. She shared that she grew up, careerwise, as a software engineer in a "group for a fairly mature product that was located in a medium-large city, which was anything but a tech hub. There were scarcely any software engineering jobs to be had in town." The result was that many people stayed at the company for a while, and she had to build her skills in an environment where she was ten to twenty years behind in terms of product and engineering experience. There was also no real way to move up, given jobs tended to go to people with long tenure. "Management positions always went to someone who had been there forever. Even lead positions were difficult to get, as these too tended to go to people who had seen the beginning of things, people who knew code that wasn't even used anymore." She also saw that women had to have an extra edge. "And among those senior peers, I never saw a woman promoted to leadership roles. There were some already there, who had been there since before I arrived, but it seemed all of them had had to arrive with that little bit extra where others didn't. The sheer pedigree of the women in leadership roles versus the men is—in retrospect—quite shocking."

Within this structure, it was also difficult to go to conferences since seniority was taken into account for those opportunities as well. She asked her boss to attend a big upcoming technical conference. "I think a Microsoft conference was around the corner—I was told that I was lucky, because there was still a slot for someone to go to the Grace Hopper conference. I was crushed. And then I was angry, because I figured out that the Grace Hopper conference was the only conference engineers in my R&D group who happened to be women ever got sent to." For context, the Grace Hopper Celebration of Women in Computing is held annually, attracting prominent speakers and thousands of attendees from across tech. Because it's specifically for women, though, Stephanie felt boxed in knowing it was the only conference she would be able to access.

However, this led to a pivotal moment in her career. "While [researching] the keynote speaker, Nora Denzel, I saw that she did 'mentor walks,' mostly for women located in Silicon Valley, which I wasn't at the time, but also via phone, if requested. She provided a career advice/mentoring session in exchange for a donation to a nonprofit in Silicon Valley [that] seeks to encourage and empower young girls interested in the tech industry." At Grace Hopper they met for an hour to talk about Stephanie's career, review her resume, and exchange advice. "She assured me that there were leadership roles that my background and experience were suited for." She also explained other roles within the technical space, and that conversation led Stephanie to get her PMP certification and transfer to program management. "It's not just that Nora Denzel gave me advice, but that someone in that big of a role, with her track record, would take their time to speak with me because she [thinks] it is important to encourage people in my role—that made an impact."

Similarly, Christine was aided by sponsors at important times. "As I think back on my career, even back in computer science, it was a bit scary to get into computer science after three years out of my five-year architecture program." She's very thankful to her first computer science professor, who believed in her and asked her to be his TA after her first class with him. This gave her a boost of confidence right when she needed it, "that I was good enough that he wanted me to help TA [for] other students. From there, because he asked me to be his TA, other professors asked me to be their TA." Even small nudges like that can lead to bigger things, and now she watches out for others. "There are many such people throughout my career that I am very thankful for in how they have sponsored me and helped me. I think that's something that I would like to do, and I would like to get more people to do, for women in tech as well, because encouragement and feedback is helpful."

DON'T LIMIT

Frances calls out that often we look for female mentors, but we should keep an open mind to others who might have had relevant experiences as well. "If you only look for people who look like you, you're going to limit yourself." She emphasizes "learning by inclusivity versus exclusivity, because I think when people are looking for a mentor who looks like us, it's actually an excluding thing. It's not an including thing, and it's not the only way to learn. It's not the only way to grow."

Kris also called out self-awareness and the benefit of different types of mentors. "Understand the things that you do that are based on the society that raised you and know that that isn't the way you have to be. Seek out other women. Seek out men that you admire attributes within and get them to teach you. . . . And I think that you can learn something from both, both the women and the men in your life, about how to navigate yourself into these meetings without compromising who you are."

Karen also encourages women to look for allies: "As Mr. Rogers would say, look for the helpers." Support could potentially be found in every corner. "I would say look for the allies wherever they are across the company, whatever kind of roles. People that you connect with or have a good feeling about or feel you can trust, because they can strengthen not just your network but they could be a sounding board for you, and they can help you navigate all this stuff that exists inside a company." She gives various examples where broad connections could help. For instance, you like your company but know your team isn't the right home. She's worked within political companies where knowing whom you could trust was critical. "You have to hold your counsel more carefully. . . . But even so, there are always some helpers and some people who can help you along the way." In the end, she says if you can't find those helpers within a company, then leave. "If everybody seems mean and out for themselves, then don't waste your time trying. You're not going to change them [and] you're not going to change that culture."

Also worth noting: some of us also find our mentors close to home. Ginny explains how important her parents were in her career. "How grateful I am for the parents that I have, because they were so wise and they were sensible and mature and offered a lot of wisdom and support." Despite both of them passing, Ginny still relies on their wisdom. "I often say this, and I mean it wholeheartedly and quite literally, that my parents have stayed with me." After her father's death she struggled, but ultimately, through mentorship, she's been able to feel their ongoing presence—"when it comes down to it, that's where I go."

COACHING OTHERS

Many women have turned to championing others as their careers have progressed, finding that it's even more rewarding to give than receive.

Gretchen provides an overall perspective on why this is important: "There continues to be a huge opportunity in tech to invest in talented people from diverse backgrounds, including women and people of color. Particularly at the entry level, I'm passionate about creating not only a more inclusive environment, but one that continues to foster talent and develop the next generation of tech leaders. There are so many opportunities in Silicon Valley for people from a range of backgrounds. Increasing the diversity of perspectives and experiences is both the right thing to do and will lead to better business outcomes."

Mentoring others can be fun too. This is worth remembering. We often feel insecure when reaching out for mentorship and over index on the time mentors are giving up, but often it benefits them too.

Sheri's always enjoyed and found success with normal relationships that turned into mentorships. "Finding the people that you want to be like and then engaging with them and creating that relationship—I think more often than not it can turn into a mentorship." Now she also mentors through both informal relationships and a formal mentoring program for tech writers. "And so I get these brand-new manager mentees and they're amazing, and it's so fun to be able to be a mentor as well. And you realize, oh wait, I have something that I can give to these people, because I've done that."

After having independently navigated her career and wishing she'd had mentors earlier, Tieisha has a more personal perspective of how mentoring and recruiting helps her. "It makes me happy to be doing well in my career, but it makes me even more happy when I know that I'm helping someone else. There's no greater feeling than knowing that you've made a difference for someone else." She talks about often being the only black woman or woman at all in her role or in the room. "I don't want to be the only; I feel a responsibility to bring people up with me." She's helped people she knows hear about other roles and opportunities because it's the right thing to do, noting it's not only about women or minorities. "It's about helping people, making a difference in other people's lives. You can fall into being selfish and just thinking about yourself, but it's way more rewarding when you can look back on your career and say, 'Wow, I actually did make some difference somewhere that wasn't to benefit me.'"

When I interviewed Marily Nika, she asked me whether she could write a section on mentorship for the book. At the time I wasn't even sure how I would incorporate it, but I immediately said yes. In the spirit of having part 3 of this book provide some practical tools for readers, Marily crafted a perfect addition with guidance, tips, and tricks for making the most out of a mentorship session. Please reference the appendix for that resource.

Amid all this advice and help, how are we actually doing? My working title for this book was, "Surviving or Thriving: Women's Stories of Building Careers in Tech." While I eventually changed it after reflecting on the spirit of career journeys, the seed of it remains in the next chapter. Do we feel like we are surviving or thriving in our jobs? And what do we do if we're struggling?

10 | Survive versus Thrive Days

It's 2017. On January 6, I received a call from my boss. I had just returned from a vacation in Mexico with my family. This was the first day he could catch me to tell me the news: There had been a leadership decision to realign the business and drastically reduce the number of team members. We previously did a round of cuts a few months prior that we all worked hard to land, but now leadership wanted a much deeper cut. I curled up in a ball on the couch. I felt like I'd been punched. My heart raced. I told my boss, "This might be it. I might be done."

Somehow I rallied. I proposed a plan to create a new organization for the hundreds of team members who needed to find new roles. I spent most of that year working with Human Resources, Staffing, and other volunteers to coach and support the individuals looking for new roles. It went well, and we placed the vast majority of team members. But I still needed to find myself a new job, and as the primary breadwinner for my family, that weighed on me.

By September I noticed I was having vision issues. It was hard for me to focus, especially when I moved. I would become dizzy, and sitting at my computer one day, I almost fainted. Then another day, while walking outside with a colleague, my left foot went numb. And stayed that way.

At that point, it had been over a year since I found out we had to do the first round of business cuts, and also a year since I experienced a family crisis at about the same time. My body was giving out after a long, long marathon.

That wasn't the first time I've come face to face with emotions and fatigue in the workplace. I have plenty of those stories, but that was the first time I thought to myself, *This might be it. I might need to quit.*

Two years later, I'm onstage in front of 1,500 program managers. I'm describing what happened to me and how I survived. How I worked through the health issues, and what I recommend for seeking balance in these often-chaotic roles. It's a presentation that resonates with many, and I receive positive feedback and invitations to speak with other audiences.

During the Q&A afterward, I repeatedly remind others of one thing: "There will never be a day when your company comes to you and says, 'Hey, close that computer. You've done enough. Go get some rest.' Work will take and take and take. It's going to be up to you to set your boundaries."

How have we survived and thrived in fast-paced environments? Where do boundaries work? Where do they fail? Surviving to thriving is really a continuum. Some women would laugh outright when I asked them whether they were surviving or thriving in their current role. "What day is it?" one asked. I interviewed women as they were changing roles, looking for new jobs, in between companies, recently promoted, having a bad day, taking care of sick children or ailing parents. The list goes on, and it all impacts how we feel in the moment.

The most common theme I noticed was that there was no permanence to this idea of thriving. There wasn't a set way you could plan your life where you could be sure to achieve it or keep it. Thriving and surviving shifted depending on the year, month, week, day, and sometimes even minute. Women would refer to the past, future, and near present in telling me how they were feeling, in a "I was just thriving" or "recently surviving" way. Some women weren't sure they'd ever feel like they were thriving, while others could be assured of feeling it as long as they were challenged or learning. Contrary to how we tend to think of people as successful or not, how we feel can be impermanent.

In my presentation about well-being, I talk about how my work stress and life stress are being navigated together each day, sometimes up and down together, sometimes at odds. The questions we personally grapple with are relevant to our work and our products:

How can we be great leaders and employees if we're struggling physically or mentally?

How can we build great products or services if we don't understand

and accept the full implications of being human? And how do we make both good work and good life decisions?

HOW WE FEEL

A lot of what we consider surviving and thriving is how we feel about what we're doing and how the day went. I've shared various stories from ladies below about what that looks like for them. First, a story from me about how this can change from moment to moment.

My kid goes to daycare relatively close by, and one day I get a call that he put Play-Doh in his ear. And they are not allowed to get it out. So I'm instantly in fixer mode—I'm like, *Okay, I'm coming. My husband's away so I'm going to come, I'm going to go over, I'm going to figure it out. Are we gonna go to the doctor? How much is this changing the day?* I get there. And they're sitting with him and they say, "You can try." I walk over and look. It's still in the outer ear. I'm like, "All right, you have to hold still." I flick it out, and all of a sudden my day is back on track and I'm totally back in thriving mode. The feeling was: I'm the best mom, and I can get back to work!

Take the below as a similar moment in time for each of these women. For this chapter, I simply give you examples of what I heard, what the women were feeling, how they were doing as real people navigating tech, and what they did to help themselves survive or thrive. This is presented as example anonymous quotations versus longer stories.

To Be Content or Not

"My question to myself is, do I want to jump back in and do something really crazy where I'm working my butt off and working long hours doing all this stuff, like a start-up thing? Or do I want to be content? And so I toy with that. And the reason [. . .] I toy with that is because I want to retire soon, which I keep forgetting. It's weird that word comes out of my mouth because I'm close to that age to be retiring too. Right? I'm in that portion of my life, or do I want to run myself ragged like I was in my twenties or thirties? And so I toy with it. . . . But I'm not great at being content. Even in my personal life I've never been great at being content."

Amid Change

"It depends on the day. And that is more specific to the fact my company was just acquired and I'm still figuring out where I 'live' in the new world. But I don't think that feeling is only relevant to M&A [merger and acquisitions], and there is a lot of M&A in tech, of course . . . when companies grow so quickly, roles and responsibilities sometimes change just as fast, and it can be confusing. It opens up great opportunities but also requires the ability to articulate and show your value over and over and over again."

Responsibility

"Oh, depends on the day. Today I'd say surviving. Because my inbox is ridiculous, but my bottom line would be thriving. And I picked thriving because I sort of stepped out on a limb and switched into [operations] from a business partner [role]. . . . So the thriving part is because my manager was very clear that my team, even though we had very little guidance, we're one of the stars. . . . I think some people are nervous about figuring it out. . . . Whereas my approach is, okay, if there's no rules and nobody's sure exactly how to do this, let's just pick a way, let's do it. And if it's wrong, then we'll fix it the next time."

Insecurity

"Current role? I'd be on the border of thriving and surviving, because I'm starting a new role, so I'm learning a lot. . . . So I feel like there's that, *Oh gosh, am I failing in this? Am I doing this?* . . . But for the most part, I feel like I can set boundaries. I feel like I have a strong team, who I can [rely] on so I don't have to do everything."

Wonder Women

"I feel like there's Wonder Woman moments where you're like, I am nailing this home and birthday party's planned. Everything looks like Pinterest [photos] and perfect, and then a presentation at work goes really well. And then there's other weeks where . . . I forget to sign up for a doctor's appointment and then at work things fall through the cracks because you just can't keep up with everything. I think that there's probably more of [surviving] than [thriving], but those Wonder Woman moments are the ones that keep me going. Like, *Okay, you can get this.*"

Guilt

"I suppose I would probably lean toward surviving. I do not feel like I am thriving. Perhaps others would disagree with that assessment, but I think I often struggle. I was talking about that inner critic that shows up, and so a lot of times I am trying to balance my perspectives on how well I am doing, what I think other people think about how well I'm doing, and all this. I also say surviving because there's this work life that I have, but also a home life and what that means. And I'm super lucky to have a partner who handles so much of what goes on at home. But I think that guilt that I have sometimes—which he does not place on me at all, I do that to myself—does become overwhelming at times, because I want what I want."

Appreciation

"I think it depends on the day. Some days I feel like my work's appreciated and everyone likes me. Not that that's important, but they liked the work that I do in a way that they feel is valuable. That's what I mean by likes me. And then I'm tackling new issues that I've never seen before, and I feel like I'm really developing. And then other days I'm like, this whole thing is going downhill. I don't know how we're going to get out of this, or this is the same problem we faced five billion times. I thought we fixed it last time. But I think actually I would say I think that I'm thriving, given that in other teams, I felt like I was not only trying to manage my work, but I was also trying to manage my management chain, who was not trying to be very helpful for me or my career or the projects I was working on or anything that I was trying to accomplish in life. So I guess it depends on how you define thriving and surviving. Compared to where I was, I'm definitely thriving."

Nonstop

"I'm definitely a bit in survival mode. I work at a pretty high-pressure, escalation-heavy environment that is a bit nonstop . . . and so I love it. I'm fascinated by the intellectual stimulation. I'm a bit addicted to the adrenaline, but always being on 24/7, [where] you never know when the next crisis is going to hit, puts me in survival mode more than thriving."

Opportunity

"It's certainly much more thriving than surviving, right? I'm not as impactful as I want to be, but I think that's a whole host of things, including the fact that I have ridiculously high standards. And that's me driving hard, right? All my bosses give me as much rope as I want, and then when I get tangled in it, they help me get untangled as opposed to strangl[ing] myself."

Learning

"I think I'm thriving. I'm learning. I'm having fun. I'm excited to come into work each day. My husband just asked me—because I was a little frustrated—'But do you think you're learning?' I'm like, 'Yes. I'm definitely learning. I'm being challenged.'"

"I feel like I'm making an impact in my organization. I'm not getting the promotions that I'm used to because I'm in a much smaller company, and my title is much bigger than other people on my team. But I feel like if I was to describe what thriving means, I'm learning, and I feel like I'm helping our company be the best. I mean, I have the ability to influence who we hire, what the compensation is, who we promote, what increases we provide, and the employee experience to create a more diverse and open and better company. So yeah, definitely think I'm thriving. I love it."

"I feel like for me, every single role I've taken, I've been very underqualified, and it's been this huge stretch opportunity. I'm super grateful to [my company]. I feel like it's a big reason [that] I never felt the need to go back to school, because I feel like in the four years I've been here, I've learned more than I learned in four years in college and more than I would learn going to business school or some other school."

State of Mind

"I was thinking of this specific example and talking to my partner about it, and he was like, 'Recently you were nervous that you weren't doing well in your role as a manager.' I stressed out a lot, and I would come home and not enjoy it, thinking I wasn't doing things perfectly so I wasn't doing them well. I was ramping up and really didn't enjoy it. I thought I was doing a terrible job. Turns out, I got really positive feedback from my team, positive feedback from my peers, from my manager. Even recently I've had people that I managed reach out to me and say that I was a great manager, and they're actually struggling with some of the new relationships they're trying to build. And so I wish I could get to a point where I had that confidence and credibility without having to get that sort of external feedback from others."

Math

"I'm thriving. I mean, if it's a numbers game, which it is, I'm doing okay. I'm doing okay. It's hard . . . I'm an achiever. My strong personality type. I wanted to check the box and get the A. . . . My mom—please publish this, please, her name is Patricia—she was in a parent-teacher conference with me [when I was a child], and my teacher was saying some concerns about my performance and my mom was like, 'Look, I don't care if she learns anything in your class as long as she gets an A.' And I think that's deep for me."

Culture

"I'm thriving; then again, there's a reason I feel this way, because the current organization has a very happy technology culture. Very empowering . . . we're not tolerating jerks that can bully people around."

150 percent

"I would say surviving would be a great description. And I think that does have to do with that notion of balance. Right now, being at the forefront [of] how technology best serves folks contributing to the disservice of people, it can take everything. It actually literally can take everything. So then the question becomes, how do you do everything you can and still have something left over? I don't even know if that's a fair question, but that's what it can feel like. There's an endless number of things to get right, so how do you get as many as you can right while still coming home and making sure that your kid practices piano, and how your daughter's navigating being a girl at eight years old and all that that entails? So I do think that it feels like you try to be 150 percent in every context and it doesn't quite work, but it does feel like something that you try."

WHAT WE DO

I'm often in survival mode, usually due to a crazy workload and parenting commitments. What helps me? I remind myself that this won't last forever. "This too shall pass" helps me keep perspective. I double down on eight hours of sleep a night because otherwise I'm grumpy and make things harder than they need to be. I write a to-do list because my brain stresses out without a list. And yes, I do add things to the list after I already did them so I can check them off! I give myself treats for the little wins so I savor the victories.

For example, I have a mason jar on my desk. It started off empty. Every time I helped someone, completed a hard task, or received kudos, I put a pretty marble inside. The clinking noise was deeply satisfying, and it helped me recognize the moment. Looking at the jar filling up gives me satisfaction on the hard days. What's your jar? What shifts how you feel? The women below describe what works for them, from taking control to asking for help.

Saying No

"It depends on the day! But I think it's a combination of things: the ability to say no (we can't say yes to everything, despite how uncomfortable it may be to say no), understanding how you show up, and setting expectations and keeping a positive perspective."

Mental Shift

"I'm also a strong believer in making a mental shift: on tough days where I feel like I barely made it through, I try to focus on three good things that happened, or three moments that gave me gratitude. By the end of that, I may not be on either side of the spectrum (not thriving nor surviving) but can tell myself, *You did the best you can, and that is good enough.*"

The Gut Check

"I'm a very open and candid person, so if there's things I can kind of call out, I'll try and call them out, because that's to me better, to surface things. But you can't do that for everything. And building a good peer network, because then I can also ask a peer. This was a good trick: asking a peer of mine like, 'Hey, what did you think about that meeting?' Or, 'Did you notice anything?' Because then I could gut check: Was it just me (and I'm biased too), or [are] other people seeing the same thing in the room?"

Taking Time

"I took a medical leave a year ago. I was so burnt out. I went to my doctor and I was like, 'I don't know what to do.' And so I took a leave for six weeks. It wasn't super long, but enough time to work with a coach, a therapist, a physical therapist (because I had a knee injury and all this stuff). And so I took that time off and it was great. I feel like I got some better perspectives and coping tools . . . I do think having the coping skills has really saved me and made me able to survive this year."

Evolving

"I would say I'm thriving. I would say I'm flourishing. I guess I finally found a role and environment and team that are quite supportive, a really good blend and application to my skills and my experience and my mindset.

"I would say I've definitely expended quite a lot of effort and resources. First it was trying to figure out how to make these certain managers happy. Then it evolved into, how do I grow my tech career? How do I actualize my potential, how do I move upward? I know a lot of what we do is figuring out how to get from point A to point B, so the amount of intel I've amassed about that, it would be crazy to not be progressing, not be getting good outcomes.

"The other nuance I guess is that it isn't taking that much. It's not like I'm commuting [a long distance]. It's not like [I'm] stressed out or working crazy hours. It's actually very intuitive and natural, which is interesting. And that's nice too, because I would say the past five years, a lot of times it had been very aggressive, intense, all encompassing, almost consuming, maybe 80 to 90 percent of my waking and sleeping energies. Where's the rest of life? And so now I'm like, 'Oh, I'm leaning out.' But I don't know if it's actually leaning out or it's just becoming more of a normal person."

Leaning on Others

"So I don't think I have the right balance. I try, but I don't try hard enough, I don't think, truthfully, to do what I need to for myself. It would be amazing to have these lives where I could do yoga in the morning or something . . . I don't know. I feel like the surviving versus thriving question is so fascinating, because I think at every stage—at every day, maybe—there's a moment where I'm like, I am in survival mode. On the way here today, I was talking to my mom who is almost eighty years old, and she was asking about something in September, and I'm like, 'I'm trying to get through today. I can't even think about what's happening in September.' So it's not that I'm in survival mode necessarily, but . . . there's so much happening at any given time that we're all just trying to catch our breaths a little bit. Sometimes a deeper breath is warranted than [at] others. But I think at any stage, all of us are both surviving and thriving.

"And the way that I get through it is by leaning on my friends and my community for support . . . I think you have to surround yourself with people that understand; whether or not they're in the same industry, whether or not they're working as many hours, that doesn't matter as long as they are appreciating you for who you are and what you do, and also the balancing act of life. That is what I think is important. If you can't relate to the people in your community or they can't relate to you, then it's not as helpful. And then you will feel very lonely and very ill-equipped too. You won't have that net."

Knowing Yourself

"My definition of thriving is probably different than Merriam-Webster, in that the thing I like to do the most is take teams and technology that is in a state—that I don't like, that I am uncomfortable, and I see that I don't enjoy—and moving it to one where I do. But then once it's in a state that I do like, it's less interesting to me. . . . Now the team is thriving, but I'm not growing. . . . I feel like I'm treading water. Whereas when things need help and I vaguely feel like I can contribute—not that I've got a handle on everything, but like I am part of the solution and it's starting to come together—that's what thriving feels like to me."

The Right Situation

"Something that's already changing that's made a huge difference in my life has been the ability to work remotely and the focus on work/life integration. As a mom of three, this has given me an amazing chance to follow my professional passions while being a good mom during a critical time in my kids' development. I am lucky to have landed a role/management team that makes this possible; however, I find that less of these opportunities exist in the next levels. I hope to see more and more realization that remote/flexible work can be impactful. . . . Even more impactful than a traditional arrangement because of the stress reduced (clear expectations, eliminating commute, etc.)."

After hearing all those thoughts from women, I'm glad I didn't name the book *Surviving or Thriving*. That title makes our lives sound too starkly oppositional, like we should or could always be one or the other. We're traveling day-to-day, and we are experiencing highs and lows all the time. Figuring out how we navigate and affect those patterns is part of our career, and life, adventure.

Speaking of that, have you ever thought about quitting your job? I have, but only seriously a few times. Mostly I imagine utopian scenarios where I am amongst the land and the animals, but nothing smells bad and the children are all well behaved. Am I alone? In the next chapter, we find out. Why do women stay or leave tech? More nuanced than the typical news articles on this subject, the answers may surprise you.

11 | The Art of Staying

As I type this, I'm soon to celebrate my eighteenth anniversary at Google and also my twentieth year in tech. When I started at Google in 2001, I had no way of knowing that I was joining a start-up that would change the world; I didn't even know that we'd launch an email product a few years later! Every year I fill out an annual company survey providing Google with anonymous information about how I feel. For many years, the survey had a question about whether I thought I would still be at the company in five years. I think I always picked the neutral option, and yet here I am.

Why have I stayed? Many people have asked me that question over the years; I've given them plenty of time to do so! Partially it's because I've had four different roles in vastly different fields at Google, and I've had the chance to pursue various ambitions, including managing and leadership. I've also birthed three children, and the working environment was flexible and allowed me to effectively work and be a parent. I also loved the people, and solving problems, and being able to play a part in changing the world. There were so many reasons.

Does that mean I never complain? Does that mean I don't want tech to change in one way or another? Anyone who knows me knows the answer. That's like asking me whether I like everything about my family. Of course not! I know them too well. Every wart, every wrinkle, every snore. The gift of family is being able to simultaneously love them but also see them for who they are. That's how I feel about tech. I'll criticize tech all day long if you want, but don't be surprised if I also defend it.

I may leave tech one day, but mainly because I'll be interested in a different life entirely, maybe one where writing and art take up a bigger piece of my time. There are certainly days where spending my time in another spreadsheet or meeting makes me imagine a different life. But if

I go, am I leaving tech like I hear many midcareer women do? Statistics are certainly sending up alarms.

I've repeatedly heard quoted, and quoted myself, a statistic from a 2016 study by the National Center for Women & Information Technology regarding technical women: "56 percent leave their organizations at the midlevel points (10–20 years) in their careers."[43] This turns out to be a number from a 2008 research report by the Center for Work-Life Policy titled "The Athena Factor: Reversing the Brain Drain in Science, Engineering, and Technology" (SET), which discusses their study of the career trajectories of women in SET and the "powerful antigens" they discovered in its culture.[44] To quote from their synopsis:

> Women in SET are marginalized by hostile macho cultures. Being the sole woman on a team or at a site can create isolation. Many women report mysterious career paths: fully 40 percent feel stalled. Systems of risk and reward in SET cultures can disadvantage women, who tend to be risk averse. Finally, SET jobs include extreme work pressures: they are unusually time intensive. Moreover, female attrition rates spike 10 years into a career. Women experience a perfect storm in their mid- to late thirties: they hit serious career hurdles precisely when family pressures intensify.

Sound familiar? It's worth noting that this original report focused on women in SET roles, which tends to leave out people in business and nontechnical roles. Surveys and research tend to focus on those roles because that's where women are the minority. The report is also now over ten years old; however, the 2014 update to the Athena study provided evidence

---

43 Ashcraft, Catherine, Brad McClain, and Elizabeth Eger. "*Women in Tech: The Facts.*" National Center for Women & Information Technology, 2016. https://www.ncwit.org/sites/default/files/resources/womenintech_facts_fullreport_05132016.pdf.

44 Hewlett, S.A., Buck Luce, C., Servon, L., Sherbin, L., Shiller, P., Sosnovich, E., & Sumberg, K. 2008. "*The Athena factor: Reversing the brain drain in science, engineering, and technology.*" Center for Work-life Policy, May 22, 2008. http://www.talentinnovation.org/publication.cfm?publication=1100.

indicating the issues are ongoing. For example, while women love their work (80 percent in the US), 32 percent say they are likely to quit within the year.[45]

So as I was interviewing women today, in various roles both technical and nontechnical, I was curious: Do women want to leave? Do they think about leaving? Do they stop growing and want to try something else? Do they leave for family? For other industries? Do they even mean to leave? After extensive interviews, I can definitely say . . . the answer is complicated.

The answer is yes and no. When I asked women if they ever thought about leaving tech, most immediately said no. Many would pause, think, and then say no. Some would say maybe yes or maybe in the future. And a few bluntly said yes. Even with that breakdown, the full story is more complex. Many go on to say they have other interests. They put a timeline on how long they could stay in tech or describe things in tech that worry them. And many have pipe dreams of escaping somewhere and doing something totally different. Is this different from how men think? Not necessarily. But since numbers are indicating that women leave tech at a higher rate than men,[46] I'm not going to spend time outlining differences and similarities. Debate on your own!

So to wrap up our journey deep into women's experiences in tech, I will focus on why we stay and why we go in this last chapter. I'll take a tour of women who are staying in tech, what they are thinking about and hoping. Then I will also look at women who left tech or are contemplating doing so, covering the whys and where they go.

---

45 Hewlett, Sylvia Ann and Laura Sherbin with Fabiola Dieudonné, Christina Fargnoli, and Catherine Fredman. 2014. "Athena Factor 2.0: Accelerating Female Talent in Science, Engineering & Technology." Center for Talent Innovation, February 1, 2014. https://www.talentinnovation.org/_private/assets/Athena-2-ExecSummFINAL-CTI.pdf.

46 Ryoo, Shane. "Why Women Leave the Tech Industry at a 45% Higher Rate Than Men." *Forbes* (via Quora), February 28, 2017. https://www.forbes.com/sites/quora/2017/02/28/why-women-leave-the-tech-industry-at-a-45-higher-rate-than-men/#3db4410d4216.

WHY WE STAY

Answering my "Have you ever considered leaving tech?" question was straightforward for some of the women because they really hadn't ever considered it. Cathy said about tech, "It's just such an interesting field. There's so many different things you can do with it." Jill shared, "I haven't thought about leaving tech because I love the creativity." Marily similarly enjoys working in tech, which for her has meant "being involved in a fast-paced, fun, fresh, and innovative environment, where you are the boss of yourself and get to work with brilliant people from all over the world. I honestly don't think it gets better than that—my days go by without me even realizing it, and that only happens if you are truly satisfied by what you do."

Camille has never thought about leaving tech. Even when she might compare jobs at other companies, it's still tech where she's focused. She's specifically fascinated by having engineers as clients and enjoys the inherent challenges in that kind of work. Karen loves the pace and the variety of people, and still loves the influence of tech. "I know it's odd to say this in these troubled times, but the values of tech overall are my values, in that the technology can make things better, has made things better for society at large. There are more ways now to make more things affordable and available to people at scale than ever before. And that includes people around the world and people who have no resources."

The fact that tech pays well was noted by multiple women, with financial security and independence being a common callout. One woman shared that she looked at nonprofit salaries, "and it was very scary how low they pay." And also, shifting industries can trigger fears about whether the move is smart for a career trajectory. Jessica knew there were options but was trying to figure out if they made sense for her. "I'm sure there are nontech companies, but it's hard to figure out if that would be a good move for your career." After a round of interviewing, she did receive an offer from a nontechnical institution. But she thought "it would be really hard to get back into tech if I made that choice, and the pay was not a salary [that worked] because I'm the breadwinner of the family, even though the lifestyle might've been better." She ultimately took a role within tech that worked better from a financial perspective.

When I spoke with Sara, she had recently left her company, but she was still committed to tech. "I don't know what it would mean to leave tech, honestly, unless it was to do a complete career departure and try to become an actress or a writer, just because tech is so prevalent throughout every industry." As she researches options, she thinks tech will be a focus point for her search, whether she seeks out jobs in a tech company or in another industry that needs technical solutions. "I think that what's going to draw a company to me, and what I feel I bring is the ability to come in and say, 'This is how technology can help us, and this is how we prepare our people to engage.'"

Ultimately, many of us feel incredibly fortunate to be where we are with the opportunities we have. There have been tough times, but overall Caragh feels "this is a privilege and an honor . . . not a lot of people get this opportunity. So even when there are hard days, I realize this is work everywhere. There's going to be drama and politics in other places too."

The Maybes

Not everyone had a clear answer, much like me. Some women paused. They had thought about it, but mostly in a dreamlike way. Reese said she would be a farmer. "I think it's natural to have that opposite fantasy. If we thought about [mine], it is to be working with the soil." If Bethanie did leave, it wouldn't be for a competitor unless it was a "spectacular opportunity." She thinks it would have to be "something totally different, like run an ice cream shop or something like it." She has a dream of "running a laundromat, because I love to fold laundry, which is a really bizarre passion of mine. I know it's the oddest thing, but it's because I can feel accomplished. I check it off the fucking list. I have beautifully folded laundry that I can put away in a drawer. I know I got something done that day."

The Worries

Belvia hasn't thought about leaving tech, but she is becoming aware of her age as she nears retirement. "I have never had a point in time in my life where I could say, 'I have had it with this, I don't want to do it anymore.'" She remembers how, when she was starting in tech years ago, her colleagues who were older were worried about being aged out. "And so it's funny, because I'd be like, 'What are you talking about? You're still young, you're youthful.' And to see them going through what they went through and it's like, 'Oh wait, no, now that's me.'" She feels secure in her current group because it skews older, but she has seen differences from team to team. Will she feel out of place or forced out at some point?

While she thinks about it, it doesn't seem like a real worry so far. For instance, she was on an internal company thread where people were talking about how they were sixty years old and recently hired. She thinks that's awesome, not just for her personal future if she wants to stay, but also because companies need the mature perspective that experience brings. "You can change some of the nuances, but the overall practices of business and how you should be running a company really tends to be the same. So having maturity, I think it gives that balance, because if you're too young, you tend to go too fast and gung ho for things, which sometimes doesn't work out."

Taking a Breather

Some don't want to leave, but they do want a break. They find the release they need and the perspective from stepping back from the world of tech, ultimately returning after a period of exploration. Yolanda has done this two times, most recently traveling for a year with her two children and husband while on a break from their jobs in Silicon Valley. Complete with homeschooling, this is an adventure she's going into with eyes wide open. She took an extended honeymoon ten years ago to backpack the world. "It was a no-brainer. One of the best decisions I've made in my life." While she had been worried back then about getting a job when she was back, now she knows she'll be fine. "I know what I'm doing; I'm very comfortable that things will be fine, and I will definitely not have regretted it."

She leverages these journeys as a reset. "I was barely [getting] six hours of sleep each night for months. I was definitely very close to burnout. I think in my life I've done things to extremes where I worked really hard and I played really hard. I find it really hard to do both." Even with that risk when she returns to work, she hasn't considered leaving tech for good yet. "It's not obvious what is better. Nothing's perfect." She thinks when she does leave, "I'll be switching to a very different mode" where money is less important than fun and personal satisfaction, like nonprofit work.

Mitali had recently left her job after fifteen years at the same company when I spoke with her. When I asked what she was doing now, she said waking up early, meditating, reading, walking, reconnecting with old friends and colleagues, and opening horizons. With degrees in engineering and business and experience in consulting, tech, finance, and human resources, she came to realize she had essentially created a general management rotation of her own with her career path to date. Where should she go from here? She felt like she needed a break to think about how she wanted to use these skills, especially because her most recent leap from business development to HR "gave me a lot of courage that I could jump into anything new" since that was a big transition she made without the relevant experience. She's taking the time off to reflect on "what I value and what's important to me before I jump into another role."

At Risk

Some women have thought about leaving for various reasons, although there are themes. According to DDI's Frontline Leader Project, 57 percent of respondents have left roles due to their managers,[47] so it's not surprising I saw this show up in the stories. Marily explains how she almost left once. "I was not motivated and did not feel that I was performing well. I spoke to HR at the company I was at, and after talking this through with them, I realized that the source of the issue I was facing was due to a manager that was not supportive." She stresses the importance of "having a manager that inspires you, that motivates you, and that provides the room you need in order to grow." She changed managers and immediately felt the difference. She sees now how vital it was that she spoke up. Mitali also recognized the importance of managers to her career happiness and success. "When I had good managers I really thrived: people who believed, let me take risks, didn't micromanage, let me make

---

47 PR Newswire. 2019. "New DDI Research: 57 Percent of Employees Quit Because of Their Boss." Press Announcement: December 9, 2019. https://www.prnewswire.com/news-releases/new-ddi-research-57-percent-of-employees-quit-because-of-their-boss-300971506.html.

mistakes and fail, willing to stretch me. The times I didn't do well was when I was boxed into a certain role or when managers thought of me in limited capabilities, e.g., fixed mindset. Those are the two times I've been disillusioned in tech."

Diane brought up the culture of the team as a risk. She remembers wanting to quit almost every day when she was in an unsupportive environment. The management chain was dominated by men, and she assumed women didn't stay for long. "Maybe this is what people do. They come out in college, they get the job that they're offered, and then they leave." It was only when she moved teams that she saw her path forward. "I was like, *I don't know if I can stay,* but then once I found a spot, then I was like, *Okay, I can manage this and figure out how to climb the ladder.*" Now she knows what a good job feels like: it's one where you can "trust the system you work in," where your management has your best interests at heart, where you feel good about what you're doing, and where your voice is heard.

Sometimes our thoughts are related to the industry or our purpose within it. Alex says she thinks about leaving tech frequently, but she did start a new role in tech recently. "Sometimes I get frustrated about the industry and how it's possible to use so much resources and energy to create something meaningless, trivial, and possibly worse for humankind. But then I remember I can change companies or teams. Other times I doubt my abilities and my 'true calling.' Was I really meant to be an engineer? Am I any good at it?" Gosia also wasn't sure whether to stay or go. After years of working in a technical area but not actually liking the work, she looked deeper into herself to find the answer. "I did years of meditation, clairvoyant training, turned into my religion, took a sabbatical, and volunteered in Africa. And any time I would ask whether I should leave, I got a no." Ultimately she realized that serving others was her calling. She began to manage and mentor others, earned coaching certifications, and also shared her writing on Medium. Recently a woman stopped her "and said she has printed one of my articles and has it on her desk . . . I don't need more than that."

Adrienne was blunt: she has other interests, and maybe those will win out over time. "I feel like there's going to be an expiration for me. And

I'm only saying that because I think there's also so many other things that I find fascinating outside of tech for me personally." She's passionate about food and nutrition, and she may want to explore that later in life. "I'll want to explore something different, and maybe my tech skills will be a foundation for that. But, I don't know that I would stay in tech forever just for the sake of it."

She recently took a new role in her company but has pondered in the past whether she needed to take a break from her career. "Do I need to put a pause on working right now to be home with the kids for a little bit? I was never like, *Oh, I want to be a stay-at-home mom* or *I want to be home when the kids are young*. I never thought that. But as I'm living it, it's hard. The funny thing is, everyone looks at me in the hallway; people come up to me and [ask], 'How do you do it? How do you do it all?' And I'm like, 'Ha ha ha' . . . I feel like I'm surviving, and there've been moments where I've thought, *I can't do this all, I need to stop and take a break*."

In the end, Adrienne took a six-week leave and enlisted a career coach. "We did a value mapping session and spent all this time going through what's important to me. I still get really excited about having a project and a problem to solve and people to work with and just doing that." The coaching helped her see that she still wanted that working environment, so she's interested now in how she can do that with balance versus leaving. "I went into it for coaching for my career, and I feel like it was more of a therapy session." Adrienne emerged with ideas about how to prioritize better. "I need to care less about my ego. I need to not worry about what people think because what I care about is my family. And so I feel like it gave me permission to say no to other social engagements. And some of it was working at night too, right? I'm not going to work at night because I want to be home and go to bed early and focus on sleep and my husband."

Laura answered definitively: yes, she's thought about leaving tech. Sometimes, however, she wonders if it's more about wanting to leave the Bay Area versus tech itself. "And for me, I'm not even sure which it is . . . would I feel this way if I were in another industry?" She boils it down to who she is: "I'm a very ambitious person that tends to oversubscribe. My husband and I joke that I would be stressed and overly ambitious even if I were a janitor or went back to living and working on the family farm."

The lack of work location flexibility was also cited as a stressor. Alena started her career in magazine publishing, gradually becoming an online content pro and moving into a managing editor position for her company's website. After nine years in tech working in a full-time remote position, she's considering leaving. "In many ways, I feel like a mom who has stepped away from the workforce completely or gone part-time, in that my promotion velocity has slowed to a crawl." She's been frustrated watching colleagues who are on-site with less experience and lower impact get promoted or take over projects. "It stings more knowing that the person who previously held my role . . . reported directly to the head of the group, while I've instead had to report to another manager with the same level of experience as I have." She was recently promoted, but it's potentially too little too late "after 2+ years of pleading and arguing for it. While I'm happy to finally receive it, the whole experience left a bitter taste in my mouth, and I felt a little deflated." She's likely to look for a job where she can no longer be a remote worker given the implications to her career, despite believing that tech companies should embrace remote work. She hopes they will "drop the outdated assumptions that distributed employees don't work as hard."

Leaving Your Role or Company

Many women talked about this idea of knowing when to leave a role or a company versus leaving tech, some having stayed longer in an earlier situation than they would now. To this point, I encourage people to think again about their boundaries. How long are we willing to stay and try to fight against a culture that doesn't work? Or a role that isn't a fit? Or a manager who doesn't believe in us? And when is it time to move on? While I'm glad that we're working to improve the worlds we're in and that we have the ability to persevere, I also want to see us take care of ourselves, to prioritize our health and happiness. Leaving may be the best thing for us, and it shouldn't be considered quitting when we do so.

When I asked Mitali how she navigated having managers who didn't support her, she admitted that the first time, "I tried to constantly prove myself. It was a little like hitting my head against the wall." Trying to stay and prove herself didn't work since the person wasn't willing to change their mind, and it "was at the detriment of my own health and values. It took me a long time to realize that my skills and experience could be more valuable elsewhere." Once she'd learned that lesson, though, she moved quickly the second time she saw the pattern repeat. "I realized early on after several interactions that my manager was looking at my abilities through a fixed mindset, and there was nothing I could do to change that. This time around I moved quickly to find a new role." She remembers how a colleague once told her, "Sometimes it's good to leave if you've been there for a while because they might have a fixed mindset of what you are from when you first started. I didn't realize that could be happening to me because I'd kept growing. Afterwards I thought I should have done this earlier, made [myself] stretch more and think differently."

Yolanda reminded us "it's okay to just move on," especially when we're in situations where it's not clear how anything will change. Leaving can actually be better than "trying to fight through that experience and thinking that that's better for you and the organization you're in." While that might be a hard decision, especially when we think that means other people win, sometimes it's key. "It's actually the environment around you, and you will be thriving much better somewhere else."

LEAVING TECH

While writing this chapter, I noticed a post to a moms-in-tech Facebook group I follow. These online communities are often the only safe and private space where women can ask sensitive questions and discuss with others in similar situations. This time a fellow mom asked how often the other women fantasize about quitting their jobs. Or did they feel excited most days? If they wanted to quit, why? I kept the link open and watched avidly as seventy responses poured in. The gist? Women love working but hate office politics, poor management, and frustrating leadership decisions. They are overwhelmed by the pressure of performing at work and at home, and they feel tired. If they aren't inspired and can afford to, they leave to find more supportive or interesting roles or companies. Some start their own meaningful projects. If they can't leave, some will fantasize about being independently wealthy or winning the lottery. And the number-one thing that would keep them in the workforce? Making flexibility a norm in the workplace, whether through work-from-home, fully remote, or flexible hours arrangements.

I know one Facebook post doesn't constitute data, but as I wrote this book, I kept seeing more on this subject. Everyone who knew I was writing a book was a potential source of information about women in tech, leaving tech, or thinking about leaving tech. I decided to proactively ask people, even strangers, via LinkedIn about whether they were leaving or considering it. A husband reached out and said the expense of the Bay Area was the number one factor for him and his wife. He described the triple stress of having a successful career in a competitive environment, trying to provide for their family in often high-cost areas like Silicon Valley, and being a good parent. An interviewee forwarded me a post from a women in tech group she follows, again with over 100 comments about why women were getting tired of tech, after an article[48] was posted on the same topic. Many responders were the only woman or the rare one on their team. Some were tired of the mansplaining, politics, and watching out for men's egos. Some were hitting ten years or twenty years in tech and were ready for a change. Some had left and were in better situations, some not. The source of the news article was also questioned—was it clickbait to discourage women from entering the field?

Even though women were only responding with short comments on Facebook, I found extremely consistent feelings as I was interviewing women in tech or on a break from tech, or those who had already left. As I listened to what these women shared with me, I decided that the statistics and articles make it sound like losing women from the workforce is inevitable, especially when childcare looms. But the recipe for their dissatisfaction is actually very clear: we lose women when they become uninspired and tired. Their love for tech stems from an ability to change the world and their lives for the better. Above all else, if tech can't help with those goals, we risk losing them.

---

48 Akhtar, Allana. 2019. "Nearly 3 in 4 women in tech have mulled leaving the field, signaling the industry still has a gender diversity problem." *Business Insider*, October 15, 2019. https://www.businessinsider.com/women-in-tech-consider-leaving-their-jobs-signaling-diversity-problem-2019-10.

Why They Go

Back in my early days at Google, I worked alongside Jennifer Lim. Jen took on our early implementation of fraud prevention in our ads business while I was dealing with policies for which ads we accepted. "It was such a fun, exciting time, and then seeing how it evolved." Jen left to be a full-time parent and eventually returned and worked on my team. I remember that Jen never clicked with the job. As she describes it, "I went back the second time, and it was such a different company. I literally felt like a robot. It felt like there was so much already established. I wasn't able to enjoy the process as much as before." As with other women I spoke with, the joy of working was in solving new and interesting problems. Jen kept asking herself, "'Am I adding value, or am I just making sure we're not adding more problems?' It was a constant battle of asking myself, 'Is it worth it?'"

After Jen left the second time, she looked for a role she could do at home while she had small children. She had heard of Stitch Fix, a company that styled women online and sent them monthly clothing shipments. Having worked in tech, she guessed they were leveraging a distributed part-time workforce, and she'd always loved fashion. She found a job description for remote stylists, passed their design quiz, and with her previous experience was off on a new adventure. "It was fast paced, exciting, and a woman-run company." She was "thrilled to be able to work with these amazing, brilliant women who wanted women to feel better about themselves. I was so drawn to that problem."

Jen was eventually promoted to styling manager. She was managing fifty remote stylists, and it was hard but fun. "I used every skill set I learned at Google in the role: new procedures, new hiring policies, [thinking about] how am I motivating my employees." When she moved to Oregon, she couldn't keep her job at Stitch Fix because they didn't have a Portland office and had a policy against employees working in states without an office.

Now it's nearly three years later, and her children are getting older. She's driven and wants a new opportunity. She loves the fast-paced nature of tech and the ability to make changes. At the same time, she's picky and doesn't want to go back to just any tech company, and she still wants the flexibility to work less than forty hours per week. There are fewer companies in Oregon and limited opportunity to explore. She regularly speaks with fellow stay-at-home parents. "We all say the same thing: we wish we could be doing something flexible. None of us want to go back [to] full-time." She points out how many have second degrees like masters or JDs, but they also have that gap on their resumes from taking care of their children. Now that the children are more self-sufficient, they wonder, "Who is going to look at me?"

Jen gives an example of a friend who is an attorney. Her friend found out her daughter was autistic and soon realized she didn't have time to help her. She worked seventy hours per week; when was there time to get her daughter an individualized education program? Get an occupational therapist? She had to quit. Another example from Jen: a woman whose husband's job requires travel 80 percent of the time. The woman considered staying in her job, but she would be paying 75 percent of her salary for someone else to shuttle her kids around to classes and other commitments. Her company didn't give her the option to work from home, so she quit. Jen's take was that 90 percent of people who leave jobs are in a similar boat. They need to leave because there isn't a way to stay.

She's been doing more research about how companies are leveraging part-time workers, and there's a big gap. I asked her whether these women would want to do repetitive, basic work or were they looking for bigger challenges. Per Jen, this is part of the problem. "Companies assume she's overqualified. They don't want to hire her to do data entry or project management, assuming she won't take the role." Talk to the moms, Jen says, and they'll say they'll do it. They want something to do so they can use their brain cells in different capacities.

Jen's starting to think about businesses that could start to leverage this untapped workforce. Many try online freelance sites, but the market is competitive, and they still might not be chosen for the opportunities. Working with companies to hire this workforce might be a win-win. This is a highly educated and motivated crew, and Jen says "they could get a job done twice as fast as someone else." In the meanwhile, Jen serves on a board for a nonprofit that specializes in giving makeovers to women in need. These women are just out of prison, or have escaped domestic violence or sex trafficking. They are referred from the state health services departments and given bags of clothes, makeup, and a haircut. Jen's also an educator assistant at her daughter's school, working with an autistic child. She brought a mom to tears when she independently researched a reward system for her autistic child. "You take that creativity from tech and can leverage it for any job."

Hamster Wheel

I ran into Lyndsay Lyle at a women's leadership dinner where she shared she was about to leave her director role and take a break. I probably leaned in too quickly to ask why, transparently collecting information for this chapter. She was happy to share that she felt like she was now on a hamster wheel, not only in tech but also with her life in the Bay Area. With the role of tech being scrutinized globally, she was also questioning what impact she was having on the world. "I'd like to gain confidence that I'm spending my time on things I care about—notably friends, family, and community, our planet and the environment, and broader social norms that are impacting our children and our political systems." She no longer felt the need to prove herself through money or her title; she was looking for a new adventure.

Elizabeth N.'s main goal is to find a forty-hour-per-week job. After spending years in demanding roles, she is "taking a step back to explore my options and find out what is the next best move for myself and my family." This is critical for her family's balance since her husband also has a job with long hours. "We can't both be putting in the hours without compromising our families' physical health, mental health, diet, sleep, quality time, etc." Interestingly, while she has a higher title than her husband, "he is compensated almost double because he works for one of the Silicon Valley big guns, and I am in the medical device/diagnostic industry, which generally does not compensate as well." Given their trajectories, it makes sense for her to be the one to step back regardless of the compensation. Also, as her parents age, she'd love to find a job where remote work is possible since they are out of the state. She wishes the United States would "adopt a contracted, thirty-five-hours-a-week work week like some countries in the EU. That would be amazing for family life."

Stella came to a similar conclusion. After eighteen years in tech at one company, she recently decided to leave. Along the way, she'd survived unwelcoming teams, imposter syndrome, and feeling like she "had to be 110 percent right to do anything." She'd already navigated her way into better opportunities and become a director, but now she was ready for a bigger change. After having two kids and birthing a new project that she grew but ultimately got shut down, it felt like a good time to reevaluate. "I don't feel like I was doing tech for the last bit. It felt like I was doing headcount, negotiations, and [performance reviews]." She was also working on diversity and inclusion efforts, which she liked, but she still wasn't satisfied. "I think it had to do with the director role—'where people meet numbers,' and I don't like talking about people as numbers. I got to the point where I was spending so much of my time not doing the things I want to be doing."

She took three months' leave to figure out the outside world before taking the plunge, having multiple coffee chats a day to probe her friends and acquaintances about their jobs. The day after she left her company, she had an idea for a children's book about technology, and now she's researching "how to explain binary numbers to a nine-year-old." Beyond that, she had a lot of options she might pursue, from consulting to building apps. "I realized talking to people that I have a lot of experience—both technical [and in] management and DEI [diversity, equity, and inclusion]—that small companies can benefit from." She does want a flexible schedule regardless of what she pursues. "Now the kids are interesting, and I want to be there. And my oldest said in May, 'Mommy, you're never here.'"

Time for Something Different

Women also naturally change focus and move on to a different career phase, shifting far from tech. Emel left to join her family's business, Betsy to psychology, and Kim to childcare. They are now embarking on full-blown second careers, and I was curious how they viewed tech now, looking back from another field.

Emel, a first-generation college student in her family, studied marketing in college. "I didn't really have anyone to show [me] the ropes

[of] navigating college. I just did what friends did." She took that skill of fending for herself and translated it to her career. She joined Google in an entry-level role and worked her way from reviewing ads to starting an office in Turkey and then Marketing. She loved the company. "I haven't grown anywhere as much as [at] Google. I haven't felt accepted and encouraged to the level I felt at Google." At the same time, she did feel out of place as an immigrant refugee. "I started to feel like these people are not like me, and I'm not like these people . . . everyone around me is privileged, from Harvard, Stanford, and Yale." She'd also married earlier, so she was in a different place than her colleagues, who were young and partying.

She ultimately left, mostly because she wanted to push herself. She moved to another tech company, but she found the culture tough there. "For fifteen years I was in a culture that promoted speaking up, doing the right things for customers. It felt the opposite there." She stayed for two years but was not happy, and then the family business opportunity provoked her to depart. She thinks she might have stayed because a leader whom she respected was joining just as she was leaving. "That's how much people in leadership matter." She currently runs a division of her family's grocery business. It inspired the part of her that loves small business, and she wanted to help grow it.

In the future, she doesn't plan to go back to tech. She feels like she sacrificed a lot in the family and work/life balance areas when she was there. She doesn't think she gave her children what she would have if she hadn't had such a demanding job. She works hard now too, and her hours aren't better, "but it's different when you make that choice for yourself because you're trying to make something successful, instead of [because] there's no path forward unless you do it this way. This industry demands it." She notes when you're younger that cost doesn't feel so high, but now she's not sure she wants to give her time like that anymore. She does miss the stability and predictability, however. "You know if you put in [the work], you will be rewarded. It's that kind of environment. One of the things I regret [is that] the financial stability is not the same." More important, she misses the people, "the thing I really value that's priceless." She was inspired by the people around her every day in tech, and she misses feeling challenged by her peers.

Betsy studied applied math as an undergraduate, earned a master's in statistics, then returned to school in 2016 for counseling psychology. After an internship made her realize she didn't want to be an actuary, she arrived in tech after starting a career in data analysis. While she could leverage her analyst skills, and she "loved tech—loved the environment, people, and pace of growth and change, the belief that hard work and innovation can lead to great things," she never really found her happy place. She often had to move on to new teams to continue to feel challenged and didn't feel connected to the products (or sometimes the egos) that went along with tech. She tried two different companies, even leaving for a time for school and travel before finding her purpose: becoming a therapist.

Even knowing her purpose, it was hard for her to take the leap and leave. "I was used to tons of email and the constant workload. For me, serving other people was fulfilling. When I didn't have that, who am I?" She also debated being a stay-at-home mom, feeling guilty that she wasn't at home more. Despite the challenges, she has been able to finish her master's degree, and she's now an intern completing her required hours at a counseling center. While happy to be on her way, she does wonder: If she'd stayed put in some of her roles and been promoted, would she have ended up finding more meaningful work along the way? She fantasizes a bit about returning to tech in a counseling capacity since she loved the environment.

Kim also made the jump into a totally different world. She left her postcollege career in tech to become a child educator. Through a combination of opportunity and her tech-built business and people experience, she's now a director of a childcare center far earlier than she expected. While she feels overworked, she's been able to experiment with her schedule so that it works better for her busy life with three kids.

At the time she left tech, it was really about the job and the company. "The role I was doing ceased to be interesting. It was the same argument with a new group of engineers every day, convincing a new group each time, no matter what I did. The company was growing, and it was not a great vibe." She was also pregnant at the time with her third child. "If I was not pregnant, maybe I would have tried . . . I don't know, but it would have affected it. Part of why I wanted to stay was the benefits, but I felt it was also hard to move to a new role when I was going on leave." She didn't see many options at the company, and moving to another tech company felt like more of the same. At the same time she was really interested in childcare, and there was a clear path to becoming a teacher and then an administrator. "And I had a super supportive husband who was 'Sure, quit your high-paying job to go back to school and then start at $17 an hour.'"

Now that she's outside of tech, she often counsels people "that if they leave, they are still a worthwhile person." She's noticed that people's whole identity can get wrapped up in where they work. "From the other side, I can say that's ridiculous." When people say it's hard to leave because of the perks, she asks, "Are those few things worth what's bothering you?" She reminds others that "there's a world outside" with jobs, restaurants, and "people with diverse experiences." She hasn't second-guessed leaving tech. "I don't regret it ever, even on the most difficult days."

Listening to these stories, I ultimately felt comforted. The publicized stories of women leaving tech due to terrible experiences can be truly awful, and I wondered if I was going to uncover masses of similar stories in my interviews. It reassured me to predominately hear stories of women evaluating and assessing their careers, not without obstacles, but in deliberate and self-focused steps. I don't want to see women running away from something; I want to see them running *toward* their goals.

This chapter reinforced an interesting message to me, though. We can't simply lay the blame for women leaving tech on bad bosses, horrible companies, or having babies. There's a daily wear and tear, amid the normal operating world of tech, that impacts them and ultimately makes them seek other opportunities. This raises a bigger question: What can we all do to help?

Conclusion

Hope can take on a life of its own. —Michelle Obama

In adventure stories, the hero is never truly alone. Whether there's a sidekick, a confidante, or a trusty crew, the hero has support and help to achieve the noble quest. In fact, the hero often must learn that we all need a team to achieve amazing heights. When I started writing this book, I wanted to share the stories of women in order to help other women. I didn't have aspirations of changing the world through grand gestures, radical policies, or revolution. Rather, I thought about those terrible moments in every adventure when we do not believe we are special enough, when we doubt our own power and abilities, when all seems in doubt. In those moments we need a nudge in the right direction, the promise that tomorrow is worth the pain we might experience today, the belief we can succeed. That's why I wrote this book, so that our stories weave together into a safety net and provide the boost when we need support and hope.

Of course, I couldn't resist seeing whether we *could* change the world.

My final question for women was, "If you could change one thing about working in tech, magic-wand style, what would that be?" There were themes: more women in leadership, more flexible work-from-home options, less arrogance in the industry, and more diversity of thought. I've included their ideas below, not only for other women but also for our allies to see. At the core, these are the ways we can attract, hire, and retain women in tech.

I'm reminded of an old game we played as children: "stiff as a feather, light as a board." It was a game we played at sleepovers. We'd surround one girl who was lying on the ground. We'd place our fingers beneath her and recite, "Stiff as a feather, light as a board" like a meditation prayer. After thirty seconds or so, we'd lift the girl using just our collective fingers. As a child it was magical. But it's not magic—it's the collective effort of all of us helping each other, a power we often underestimate. And when we all do it, the load is light.

HIRE A DIVERSE WORLD

"More women and more minorities in higher places. Not just women, but it's women and minorities. If you look at the higher [levels], they've all pretty much had the same kind of backgrounds. We tend to gravitate to what we know. And so if you don't have people that are of a different race, color, gender, or whatever mixed in it, you're always going to have what you know, and you're never going to see it differently."

"I feel incredibly grateful that in spite of my lack of formal education, I got to work for the best company in the world to work for. I see more and more of that happening and coming and giving opportunities to people that didn't necessarily go down the most conventional path. I think that some of that is because of some of the forefathers of tech, right? Like Steve Jobs, who didn't have a formal education but was creative and a critical thinker and really saw a vision. I think that out of any industry, tech is really the forefront of seeing people for their skill and their mind and allowing them to come and have a seat at the table. I hope that we see more of that, because there's a lot of really incredibly brilliant and inspiring and passionate people out there that could have such a huge impact on our world and society if we just give them a chance and a seat at the table."

"I would make all the management teams 51 percent women . . . I think it would help with pay equity. I think it would help promoting women through the ranks, and I think it would help generally building more diverse teams."

"That we blindfold ourselves when we are looking to recruit talent so that we have organizations that are a bit more reflective of the real world. I say that because in my career, I've only worked for big companies, and the power structure has always looked the same. It's never changed."

MORE HUMILITY

"It has to do with entitlement, because I see it across the board . . . I think it would be if everybody had to have worked as a waiter in a restaurant for two years of their lives. Like some *real service*. You're not allowed to start . . . if you have not had a regular on-your-feet, people-may-or-may-not-treat-you-well job for two years where you have to bring your own lunch or pay for your own lunch. Two solid working years of experience doing anything else but tech."

"I would change the myth [that] there's these unique individuals who are ten times better than anyone else, and if we could only find them, promote them, and get everything out of their way, they could solve all the problems. I think that that's a myth that creates incentive structures and promotion structures and things like that but also creates leniency for assholes. It also means that people who are like, 'Well, that's not me,' feel like then they shouldn't join because they're not one of these people. And I think that that myth has done a lot of second-order cultural harm on tech . . . I would make that myth disappear, and I think that over time it would have a lot of positive follow-up in a lot of different areas."

"I would change the mindsets of a lot of the leaders to have them appreciate differences and have them have a sense of humility, because I think we are crippled by our own success in some ways. We've got big blind spots that I think are going to come roost in the next ten years if we don't find some humility and self-awareness . . . and I think only strong leaders can help infiltrate and crack open some of that exclusivity that we have unintentionally allowed to make us rather myopic."

MORE TRAINING AND CAREER DEVELOPMENT

"While it's often different at every start-up or tech company, at this point in time I would say [that] within HR and career development, I don't think enough value and prioritization is placed upon a structured and thoughtful setup for both of these areas. For women and other minorities, this can impact our careers from the moment we are hired—and we may not even realize it until later."

"Please offer nontech people technical university. The technical roles are what pay better and what lead to the ability for a businessperson to move more adeptly in the marketplace."

"You have to really much more develop and extend your [learning and development] offering for the full spectrum of people. Not just the thousands of young graduates you're bringing in, but the tens and twenties of your valued older employees. And don't make it so entrepreneurial. Look after people throughout the course of their career with you. And beyond, by the way, because that alumni network really is valuable. It makes a big difference."

BRING EVERYONE ALONG

"All the clubs for [women or other affinity groups] . . . I would invite them to be more inclusive. Bring men to the women gatherings as listeners or guest speakers. Maybe 10 percent of the audience could be male? Ask them to give us scenarios where we can #leanin better, or be bolder, or be braver. A woman needs to operate in a man's world. That is the reality of today. How can we learn skills to do so better?"

"There would be more of a partnership across businesspeople and engineering, as opposed to what I would say is the tendency toward engineering dominance and everything else traveling behind."

"I think it's more mentoring and sponsoring of women midcareer. I think there's a reason they talk about a dip at [midcareer]. Women don't know how to navigate it. And tech is nebulous [about] what keeps you going in your career. And in some ways it's painted as meritocratic, but tech is no different than other industries—it gets narrower and narrower, [and] there's only so many positions available to you. Women need some guidance along the way about how to be more deliverable in managing their careers, counseling help and coaching along the way. When I think back on my career, a lot of my success has been because of the male managers and mentors that I had along the way who believed in me. When I expressed desires, they believed and told me to go do it. I attribute much success to those silent folks in the background. I wish there was more of an effort to recognize that and training on how to do that."

A FLEX WORLD

"Flex hours and flex location. It's strange to me that in this very wired world, most employers . . . expect you to be on-site at all times. Work from home is a rarity. I don't understand how people have a problem with that. If I was a boss of people, I would not care. [They could be] anywhere—that can be on the beach, they'd be hanging upside down and skydiving. As long as I could reach them and they got their work done, they're there when they are needed to be there for presentations or a meeting, I don't care. They want to work from 3:00 to 5:00 a.m., that's fine. I feel like you need to have a level of trust between you and your employees, and understand when you have hired them that they're capable of doing what they say they can do."

A REAL MERITOCRACY

"So I think magically, if we could take away any unconscious bias and it was really merit based. In the absence of that, if anyone [could] sit me down and explain, 'You didn't get this role because of X.' . . . If you want that type of job in the future, how are you going to be able to figure out what it's going to take to get there?"

"It would be true equality. So when I look [at] whether or not it's tech or our societal environment at large, the amount of talent and passion that we waste by treating people unequally, or by these unconscious biases that don't let people fulfill their full abilities. I think it's tremendous. If you start one of these *Star Trek* movies and they go forward into the future, you have these extremely diverse crews of ships, and you can't predict who's going to be the scientist or the captain or things like that. That seems like the right kind of future, not just for tech, but for everybody. It isn't about race and gender alone. For me, it's also deeply about socioeconomic status. I would love for the children of coal miners to be represented at Google and the children of soldiers to be represented. And so that notion that talent can and should come from everywhere and anywhere, that notion of a world in which we could get that right, is in some ways my personal Shangri-La. I think that would be amazing. Granted, we're only somewhere along that path, and hopefully we get there. But I can't imagine the kinds of things that we would already have achieved if we had a true appreciation of human capital wherever it is and whoever it is."

Epilogue

As I was wrapping up my first draft of this book for copyediting, COVID-19 was swiftly on the rise. I was sitting in a cafe at a Target waiting for my phone's screen to be repaired across the street. Racing to edit the book together, I saw cancellation emails pouring in, most importantly the closing of my children's elementary school but also events like conferences and birthday parties. The Target was already stripped clean of hand sanitizer and toilet paper. I wandered the store trying to imagine how I would keep my children busy, grabbing art supplies and construction toys. Knowing I would work from home and suspecting that school was done for the school year (if not longer), I silently accepted that my old life, and its pros and cons, was gone. The world was changing, and it was hard to grasp what would replace it.

Humans are pretty bad at imagining a different future state from the current reality. This is left over from spending many years simply trying to survive. We are good under sudden attack like running from a bear; our built-in instincts take over to guide us. Invisible enemies or whole-scale change is hard to prepare for, though, and instinct alone doesn't help us suddenly own our children's home education. I suppose that's why I can't say how the world will change from all of this, but I can say it will. Even watching my extroverted, energetic four-year-old deal with sudden social isolation, I know this is leaving its mark.

As a lifelong introvert, this new world seems built for me. While I know it won't last, the freedom of not having to do *everything* may stick. Another thing? I've watched tech accept that this is a major global event, and it's not business as usual. People are grappling with significant emotional, financial, and physical upheaval, and tech has accepted that in a way I've never seen previously, broadly focusing on our health over productivity for the first time I can remember. And as we see people's families and homes on video conferences (VCs), it's clear we're all people coping the best we can together.

The dark side? Many women still feel like they are losing more as the costs of the pandemic take their toll. Still bearing a large part of the education and domestic burden given our role as caretakers in society, this period of time threatens to further hold back our personal development as we work to take care of children, parents, and friends. On social media, new career questions are being raised, like how much to show an infant on VCs. Is it unprofessional, distracting, or a welcome respite? Is this an example of social progress, or will it ultimately hold us back? And as women are laid off in this troubled economy, they wonder how they'll find another role to provide for their family and sustain their career trajectory. As stark questions are raised about our society and who is blessed and who is burdened, it's hard to predict what will change for the positive versus the negative.

And so, the questions I wonder for tech are staying the same, but with even more urgency. Can we hire a diverse workforce, so we build products that meet the world's needs? Can we build products that will invest in the criticality of core survival services as much as the latest high-end gadget? In a world of haves and have-nots, how can tech help narrow that gap instead of exacerbating it? With so many of us working from home, can we invest in a different way of working in the future? A future where we don't need to be physically together to align and build products? And in that world, could we then hire even more diversity from across the globe? And after we do all that, will more women thrive in tech?

I wish I had more answers. I'm still dealing with my own questions of how I'll keep the kids busy this summer and how to do a book tour without actually going anywhere. But I'm excited. Never has the tech environment been so ripe for change. For the first time, I speak regularly with my boss about how my team is feeling versus the latest projects and escalations. Companies are announcing long-term flexible working arrangements as they see work-from-home arrangements succeed for their organizations. Tech has shifted focus to provide health resources and information during this critical time, instead of chugging on their own plans.

Pivots like that stand the chance to benefit us long after the COVID-19 threat dissipates. I hope that this book can serve as a pre-pandemic time capsule, a relic of the good, bad, and ugly, as we continue to evolve toward a better future.

Now I've got to go, and take the kids out for a walk in the sunshine. Take care, and thank you so much for reading. Please share with your colleagues and friends. Comments and feedback welcome to adventuresofwomenintech@gmail.com.

Acknowledgments

Before writing a book, I always wondered why acknowledgements sections in books were so long. That's until I asked everyone (and in one case literally their mother!) to help me with this book. So a huge thank-you to my village. I could not have done this without each and every one of you.

First off, a huge thank-you to all the women I interviewed for this book. Thank you for allowing me into your busy lives and careers. I am forever indebted to your generosity, and I have learned so much from each of you.

To Carolyn Stephen, Georgia Dealey, Jennifer Tacheff, Michelle Peterson, Rachel Hyman, and S.M. for being my cheerleaders, advisors, and confidantes. I'm sure I seemed like a crazy lady at times, and you never told me so. Bless your hearts.

To Kate Brodock, Jonathan Rosenberg, and Sheryl Sandberg for taking a bet on me and making me believe I could succeed. You inspired me, not just to write this book, but also to help others.

To Karen Wickre for her endless advice, connections, and support. I'm so glad I emailed out of the blue and reinvigorated our loose connection.

To Marily Nika for not only answering my questions but also asking whether she could contribute to the book. Your experience and insight are appreciated.

To all my early readers, including Alicia Ling, Arpana Tiwari, Camie Hackson, Christine Chau, Dmitry Lazarev, Erica Lee, Fleur Knowsley, Genevieve Strycharz, Grant Rose, Jacob Brace, Jessica Loss, Joseph Little, Lyndsay Lyle, Maya Razon, Renée Richardson Gosline, and Sherry Lin. Thank you for reading not-yet-done chapters, for your feedback, and for your encouragement.

To the colleagues, acquaintances, and near strangers who encouraged, helped, or advised me along the way, including Adam Smiley Powalsky, Alan Eagle, Alison Bloomfield Meyer, Brian Kernighan, Carlye Greene, Greg McBeth, Hans Peter Brondmo, James Levine, Jennye Garibaldi, Liz Dubelman, Manita Sharma, Marisella Bodrero, Mitch Joel, Raena Saddler Schellinger, Rebeca Hwang, Robert K. Roskoph, and Samantha Karlin. You didn't have to take the time, but it made all the difference to me that you did.

To my wonderful partners at Wise Ink, especially Dara Beevas and Patrick Mahoney, for bringing my book to life. I was lucky to find you. And our partners: Anitra Budd, Katharine Bolin, Luke Bird, Marshall Davis, and Tom Stoneman.

To the endless list of my coaches, advisors, sponsors and mentors: I was listening even when I didn't seem to be. Thank you for your wisdom, patience, and time.

To my past and current team members who've taught me so much about the leader and human I want to be, thank you for inspiring me every day and letting me practice on you.

To my mom, dad, and brother, who never tolerated me being anything less than I could be.

And finally to my husband and children for their endless support of this book and my work, particularly on my grumpy days.

APPENDIX

Mentorship 101

By Marily Nika

TL;DR: Mentoring is a fun and adventurous experience that will always leave you with a story to tell.

WHAT IS MENTORING AND WHO IS IT FOR?

There are many definitions on mentorship online. For me, mentorship is being able to learn from someone in the same field as you, that you truly look up to. Here are some more definitions:

> Mentoring is a system of semistructured guidance whereby one person shares their knowledge, skills, and experience to assist others to progress in their own lives and careers. —University of Cambridge

> Mentorship is a relationship in which a more experienced or more knowledgeable person helps to guide a less experienced or less knowledgeable person. —Wikipedia

The way I see it, there is no one-size-fits-all definition of mentorship, and that's exactly why it is for everyone. Mentorship occurs when two people who share a passion for a certain profession get together to discuss their experiences and support each other for professional growth, success, and confidence.

Mentors and mentees can come from any background and seniority level—in fact, the more diverse the experiences, the better! Thanks to the internet, you can be based in different continents and time zones and still make the most out of mentorship.

People engaging in mentorship can learn from each other at the same time *(reciprocal mentoring)*, learn together *(peer mentoring)*, or have the more junior one teach the more senior one. At the end of the day, mentorship is what you make of it.

In tech, mentorship is more often *reciprocal*, a two-way street. That's because tech is such a fast-paced area—new technologies come out almost daily, more and more tools and resources become available as we speak, there is constant trial and error, and often it's the more junior person who will have the most up-to-date insights and knowledge that they can pass on to the mentor. The result of it is a rewarding two-way learning experience.

FINDING A MENTOR

There are many ways to find a mentor. When I was a student, I had formed a "*pantheon*" of people I truly looked up to and that I hoped to one day learn from. Even to this day, I keep adding people to this list, and I was surprised as to how many people from this list I was actually able to talk to. Here are some ways to find mentors and create your own *pantheon*:

Universities

If you are still in college or university, reach out to your career office—there most likely is a formal mentorship program where you can sign up. If there isn't one, create it!

Conferences

If you attend an academic or professional conference, make sure to use *every* opportunity for networking. It's the best way to meet people who share your interests and passions. The networking sessions are made just for this purpose: to connect!

Don't be afraid to walk up to someone and introduce yourself, hand them your business card, and connect online. You never know what connecting could lead to, from a future cofounder to a good friend or to someone who can introduce you to someone else.

Online

Don't be hesitant to reach out to people whose careers you look up to. Search on Medium, GitHub, Twitter, LinkedIn, and so on. Send them a message explaining why you are reaching out. Here's an example:

> Dear X,
> I am reaching out to you after [listening to your talk at Conference XYZ, reading your article at *Z*, and so on]. I'd love to grab a few minutes of your time and learn more about [topic].

Even if you don't get an answer, there is nothing to lose!

WHY BE A MENTOR IN THE FIRST PLACE?

> Women who are mentored by women feel more supported and are often more satisfied with their career.
> —LeanIn Tips

Anyone can be a mentor and/or a mentee, at their own capacity and pace. In fact, you may already be a mentor and not realize it if you often answer career questions or support younger people, and you most likely have been a mentee if you have reached out to someone more senior than you for some form of advice.

As far as mentoring is concerned, it is extremely rewarding. In fact:

1. Someone's life can be positively transformed by your own past experiences.
2. You can reinforce your own skills and knowledge.
3. You can keep up with what's going on outside your company/role.

4. It's rewarding in more ways than you can expect, and it can (and will!) transform not just the mentee's life but yours as well.

The takeaway is that thirty minutes can truly change someone's life, and it can change yours too.

FINDING A MENTEE

Finding a mentee is easier than you think, as the mentee will often be the one to find you.

- Social media: Read your LinkedIn messages, check your Twitter mentions, and engage with the online tech community, as you most likely will get a question you will want to answer and meet someone who wants to learn from you.

- Network at conferences and be open to conversations at your company's café or break area.

- Get involved with tech conferences and recruiting events as a speaker or panelist.

All you have to do in order to become a mentor is be open to it.

There is no standard format for mentorship, so if you feel you are too busy to mentor, here are some ways to fit it in your schedule:

- Suggest a quick coffee break or say yes to a thirty-minute coffee break when someone asks, even if it's only once per week.

- While commuting, instead of listening to music or a podcast, have a call.

- Mentor indirectly by writing articles on Medium, LinkedIn, or similar online forums about what you wished you knew when you were starting your career.

BEST PRACTICES DURING A MENTORING SESSION

- Listening is key:
    - ◊ You both most likely have limited time, so make the most out of it—truly listen to each other and make sure to communicate what your goals are and what problems need to be solved.
- Don't be afraid to open up and communicate your concerns, stories about failure, or moments of weakness—a mentor/mentee relationship is based on trust, confidentiality, and respect. The more *real* you are, the more you will get out of this session.
- Do not sugarcoat anything.
    - ◊ Failing is all about getting back up and trying again. Share your failures and your learnings, and don't sugarcoat your experiences.
    - ◊ Share your problems and the obstacles that you overcame along the way.
- Celebrate your successes, don't hide them—you got a seat at the table because you deserve it.
- Have fun! Mentorship is a fun experience that can create long-lasting relationships.

Harassment Resources and Support

The US Equal Employment Opportunity Commission provides a clear list of steps for those who have a harassment complaint.[49] While they are US focused, I have referenced and summarized those steps below. Note that laws will vary by country. For the latest US information and additional resources, please reference www.eeoc.gov directly.

1) If you feel comfortable doing so, **tell the person who is harassing you to stop.**

2) If you do not feel comfortable confronting the harasser directly, or if the behavior does not stop, follow the steps below:

Check to see if your employer has an anti-harassment policy. This may be on the employer's website or in the employee handbook. If not, ask any supervisor (it does not have to be your own) or someone in Human Resources (if your employer has an HR department) whether there is an anti-harassment policy and for a copy.

If there is a policy, follow the steps in the policy. The policy should give you various options for reporting the harassment, including the option of filing a complaint.

If there is no policy, talk with a supervisor. You can talk with your own supervisor, the supervisor of the person who is harassing you, or any supervisor in the organization. Explain what has happened and ask for that person's help in getting the behavior to stop.

The law protects you from retaliation (punishment) for complaining about harassment. You have a right to report harassment, participate in a

---

49 US Equal Employment Opportunity Commision. 2020. "What You Should Know: What to Do if you Believe you have been Harassed at Work." Accessed May 23, 2020. https://www.eeoc.gov/laws/guidance/what-you-should-know-what-do-if-you-believe-you-have-been-harassed-work.

harassment investigation or lawsuit, or oppose harassment, without being retaliated against for doing so.

You always have an option of filing a charge of discrimination with the EEOC to complain about the harassment. There are specific time limits for filing a charge (180 or 300 days, depending on where you work), so contact EEOC promptly.[50] You can also meet with EEOC to discuss your situation and your options. This conversation is confidential. Note: federal employees and job applicants have a different complaint process and different time limits.

I also want to recognize the emotional and psychological toll of harassment, which can be a source of severe stress. To decide whether to take any of the steps above, you may need support and counseling. Needing help is perfectly normal. If you need additional support, please consider[51]:

Employee Assistance Program (EAP): A confidential, 24/7 workplace benefit paid for by many employers that provides various services such as short-term intervention, counseling, and resources. EAP counselors can help navigate options and make a plan. To see if an EAP is available, refer to your benefits package.

EmpowerWork (empowerwork.org): A US nonprofit, confidential texting service to give employees a safe way to report workplace issues. It allows employees to report harassment, bullying, intimidation, discrimination, and workers' compensation fraud. Counselors have been trained and coached to both provide emotional support and distinguish between unlawful and unfair workplace issues.

Your friends and loved ones: Line up your support team regardless of what path you choose, and take care of yourself first and foremost.

---

50 US Equal Employment Opportunity Commision. 2020. "How to File a Charge of Employment Discrimination." Accessed May 23, 2020. https://www.eeoc.gov/how-file-charge-employment-discrimination.

51 Kurter, Heidi Lynne. 2020. "5 Powerful Apps And Resources To Tackle Workplace Bullying And Harassment." *Forbes*, January 16, 2020. https://www.forbes.com/sites/heidilynnekurter/2020/01/16/5-powerful-apps-and-resources-to-tackle-workplace-bullying-and-harassment/.